AU005606 TK

MW01265416

AMD/AUSTIN LIBRARY **WITHDRAWN**
Advanced Micro Devices   AMD/Austin Library
5204 East Ben White Blvd.   August 2010
Austin, Texas 78741

VHDL MODELING
FOR
DIGITAL DESIGN SYNTHESIS

VHDL MODELING
FOR
DIGITAL DESIGN SYNTHESIS

by

Yu-Chin Hsu
Kevin F. Tsai
Jessie T. Liu
Eric S. Lin

University of California, Riverside

AMD/AUSTIN LIBRARY
Advanced Micro Devices
5204 East Ben White Blvd.
Austin, Texas 78741

KLUWER ACADEMIC PUBLISHERS
Boston / Dordrecht / London

Distributors for North America:
Kluwer Academic Publishers
101 Philip Drive
Assinippi Park
Norwell, Massachusetts 02061 USA

Distributors for all other countries:
Kluwer Academic Publishers Group
Distribution Centre
Post Office Box 322
3300 AH Dordrecht, THE NETHERLANDS

Library of Congress Cataloging-in-Publication Data

A C.I.P. Catalogue record for this book is available
from the Library of Congress.

(Appendix B.1 and B.2) Reprinted from IEEE Std 1076-1993 IEEE Standard VHDL Reference Manual, Copyright© 1944 by The Institute of Electrical and Electronics Engineers, Inc. The IEEE disclaims any responsibility or Liability resulting from the placement and use in this publication. Information is reprinted with the permission of the IEEE.

(Appendix B.3) Reprinted from IEEE Std 1164-1993 IEEE Standard Multivalue Logic System for VHDL Model Interoperability (Std_logic_1164), Copyright© 1993 by The Institute of Electrical and Electronics Engineers, Inc. The IEEE disclaims any responsibility or liability resulting from the placement and use in this publication. Information is reprinted with the permission of the IEEE.

(Appendix B.4) Reprinted from IEEE Draft Std P1076.3 IEEE Draft Standard VHDL Synthesis Package, Copyright© 1995 by The Institute of Electrical and Electronics Engineers, Inc. This information represents a portion of the IEEE Draft Standard and is unapproved and subject to change. Use of the information contained in the unapproved draft is at your own risk. The IEEE disclaims any responsibility or liability resulting from the placement and use in this publication. Information is reprinted with the permission of the IEEE.

Copyright © 1995 by Kluwer Academic Publishers

All rights reserved. No part of this publication may be reproduced, stored in a retrieval system or transmitted in any form or by any means, mechanical, photo-copying, recording, or otherwise, without the prior written permission of the publisher, Kluwer Academic Publishers, 101 Philip Drive, Assinippi Park, Norwell, Massachusetts 02061

Printed on acid-free paper.

Printed in the United States of America

SUBJECTS:
DIGITAL ELECTRONICS — COMPUTER SIMULATION
VHDL (COMPUTER HARDWARE DESCRIPTION LANGUAGE)
COMPUTER AIDED DESIGN
ISBN 0792395972

CONTENTS

LIST OF FIGURES		xi
PREFACE		xv
1	**INTRODUCTION**	1
	1.1 Design Process	1
	1.2 Levels of Abstraction	2
	1.3 Design Tools	5
	1.4 VHSIC Hardware Description Languages	7
	1.5 Simulation	10
	1.6 Synthesis	11
	1.7 Summary	13
2	**BASIC STRUCTURES IN VHDL**	15
	2.1 Entity Declarations	16
	2.2 Architectures	19
	2.2.1 Behavioral style	21
	2.2.2 Dataflow style architecture	24
	2.2.3 Structural style architecture	25
	2.3 Packages	29
	2.4 Configurations	30
	2.5 Design Libraries	31
	2.6 Summary	33
3	**TYPES, OPERATORS AND EXPRESSIONS**	35
	3.1 Data Objects	35
	3.2 Data Types	36

		3.2.1	Enumeration Types	37
		3.2.2	Integer Types	38
		3.2.3	Predefined VHDL Data Types	38
		3.2.4	Array Types	39
		3.2.5	Record Types	41
		3.2.6	STD_LOGIC Data Types	42
		3.2.7	SIGNED and UNSIGNED Data Types	43
		3.2.8	Subtypes	44
	3.3	Operators		44
		3.3.1	Logical Operators	45
		3.3.2	Relational Operators	45
		3.3.3	Adding Operators	46
		3.3.4	Sign Operators	47
		3.3.5	Multiplying Operators	47
	3.4	Operands		47
		3.4.1	Literals	48
		3.4.2	Identifiers	49
		3.4.3	Indexed Names	49
		3.4.4	Slice Names and Aliases	50
		3.4.5	Attributes Names	50
		3.4.6	Aggregates	51
		3.4.7	Qualified Expressions	52
		3.4.8	Type Conversions	53
	3.5	Summary		53
4	**SEQUENTIAL STATEMENTS**			**57**
	4.1	Variable Assignment Statements		58
	4.2	Signal Assignment Statements		59
	4.3	If Statements		62
	4.4	Case Statements		63
	4.5	Null Statements		65
	4.6	Assertion Statements		65
	4.7	Loop Statements		66
	4.8	Next Statements		69
	4.9	Exit Statements		69
	4.10	Wait Statements		70

	4.11	Procedure Calls	72
	4.12	Return Statements	73
	4.13	Summary	73
5	**CONCURRENT STATEMENTS**		75
	5.1	Process Statements	76
	5.2	Concurrent Signal Assignments	79
	5.3	Conditional Signal Assignments	80
	5.4	Selected Signal Assignments	82
	5.5	Block Statements	83
	5.6	Concurrent Procedure Calls	84
	5.7	Concurrent Assertion Statements	85
	5.8	Summary	85
6	**SUBPROGRAMS AND PACKAGES**		89
	6.1	Subprograms	89
	6.2	Packages	95
		6.2.1 IEEE Standard Logic Package	98
		6.2.2 IEEE Standard Synthesis Packages	100
	6.3	Summary	101
7	**MODELING AT THE STRUCTURAL LEVEL**		105
	7.1	Component Declarations	105
	7.2	Component Instantiations	106
	7.3	Generate Statements	109
	7.4	Default Bindings	112
	7.5	Configuration Specifications	115
	7.6	Configuration Declarations	117
	7.7	Modeling a Test Bench	120
	7.8	Summary	125
8	**MODELING AT THE RT LEVEL**		129
	8.1	Combinational Logic	130
	8.2	Latches	131
	8.3	Designs with Two Phase Clocks	134
	8.4	Flip-Flops	135

	8.5	Synchronous Sets And Resets	137
	8.6	Asynchronous Sets And Resets	139
	8.7	VHDL Templates for RTL circuits	141
	8.8	Registers	144
	8.9	Asynchronous Counters	148
	8.10	Synchronous Counters	149
	8.11	Tri-State Buffers	150
	8.12	Busses	153
	8.13	Netlist of RTL Components	155
	8.14	Summary	156
9	**MODELING AT THE FSMD LEVEL**		163
	9.1	Moore Machines	164
	9.2	Asynchronous Mealy Machines	167
	9.3	Synchronous Mealy Machines	171
	9.4	Separation of FSM and Datapath	173
	9.5	An FSM with a Datapath (FSMD)	177
	9.6	Communicating FSMs	183
	9.7	Summary	187
10	**MODELING AT THE ALGORITHMIC LEVEL**		191
	10.1	Process and Architecture	192
	10.2	Wait Statements	194
	10.3	Synchronous Reset	196
	10.4	Asynchronous Reset	200
	10.5	Registers and Counters	202
	10.6	Simple Sequential Circuits	204
	10.7	Algorithms	205
	10.8	Process Communication	213
		10.8.1 Two Way Handshaking Communication	215
		10.8.2 One Way Handshaking Communication	217
	10.9	Summary	220
11	**MEMORIES**		227
	11.1	Memory Read/Write at the RT Level	229
	11.2	Memory Inference at the Algorithmic Level	235

	11.3 Summary	238
12	**VHDL SYNTHESIS**	**243**
	12.1 VHDL Design Descriptions	245
	12.1.1 Algorithmic description	246
	12.1.2 FSMD description	248
	12.1.3 Register transfer description	249
	12.1.4 Gate level description	249
	12.2 Constraints	250
	12.3 Technology Library	250
	12.4 Delay Calculation	254
	12.5 The Synthesis Tool	256
	12.6 Design Space Exploration	258
	12.7 Synthesis Directives	265
	12.7.1 Synthesis Off and On Directive	265
	12.7.2 Asynchronous Set/Reset Directives	267
	12.7.3 Function Directives	269
	12.7.4 State Variable Directives	270
	12.7.5 Don't Care Value Directives	272
	12.7.6 Register Array Directives	274
	12.8 Summary	274
13	**WRITING EFFICIENT VHDL DESCRIPTIONS**	**279**
	13.1 Software to Hardware Mapping	280
	13.2 Variables and Signals	282
	13.3 Using minimum bit width	283
	13.4 Using effective algorithms	285
	13.5 Sharing complex operators using module functions	287
	13.6 Specifying don't care conditions	288
	13.7 Writing low level code	289
	13.8 Summary	290
14	**PRACTICING DESIGNS**	**295**
	14.1 Bit Clock Generator	295
	14.2 Traffic Light Controller	297
	14.3 Vending Machine	301

14.4 Black Jack Dealer Machine	305
14.4.1 Black Jack Dealer Design	305
14.4.2 Testbench Design	310
14.5 Designing a Stack Computer	312
REFERENCES	323
A RESERVED WORDS	327
B STANDARD LIBRARY PACKAGES	329
B.1 The STANDARD Package	329
B.2 The TEXTIO Package	330
B.3 The Standard Logic Package	332
B.4 The Standard Synthesis Packages	337
B.4.1 NUMERIC_BIT	337
B.4.2 NUMERIC_STD	343
INDEX	353

LIST OF FIGURES

Chapter 1

1.1	The Digital Design Process.	3
1.2	The Y-chart.	4
1.3	Design tools.	6
1.4	Comparator (a) Entity (b) Behavior style (c) Data flow style (d) Structural style.	9
1.5	Simulation (a) hardware model (b) simulation waveform.	10
1.6	Synthesis (a) Behavior (b) RTL (c) Logic.	12

Chapter 2

2.1	Entity declarations (a) one-bit adder (b) four-bit adder.	18
2.2	Entity and architectures.	20
2.3	Behavior Model (a) Process model (b) simulation cycle.	23
2.4	Structural Decomposition (a) A Design Hierarchy (b) Design Tree.	27
2.5	Design Libraries.	32

Chapter 6

6.1	Behavior Model (a) Resolution function (b) Three hardware implementations.	93

Chapter 7

7.1	Full adder Design (a) Interface (b) Architecture (c) A Design Hierarchy.	107
7.2	Generated Adder (a) for-scheme (b) if-scheme.	111
7.3	A test bench for the full adder.	122

Chapter 8

8.1	A representation of an RTL design.	129
8.2	Two different synthesized results.	131
8.3	A simple latch.	132
8.4	A latch with asynchronous reset.	134
8.5	A digital design using two-phase clock.	134
8.6	A D-type flip-flop.	136
8.7	Reset (a) A D-type flip-flop with synchronous reset. (b) A D-type flip-flop with asynchronous reset.	138
8.8	Two implementations of an enabled D-type flip-flop.	140
8.9	Templates of an RTL process (a) A synchronous section (b) a synchronous section with with asynchronous inputs (c) a combinational section.	143
8.10	A gate level schematic.	144
8.11	A four-bit register.	145
8.12	A four-bit register with parallel load input.	146
8.13	A shift register.	147
8.14	A 4-bit binary ripple counter.	148
8.15	A tri-state gate.	151
8.16	A design contains a tri-state gate.	152
8.17	A bus system.	154
8.18	A data path.	155

Chapter 9

9.1	Interaction of a Controller and a datapath.	164
9.2	A block diagram of a Moore machine.	165
9.3	A state transition diagram of a Moore machine.	165
9.4	A block diagram of a Mealy machine.	168
9.5	A state transition diagram of a Mealy machine.	168
9.6	A block diagram of a Synchronous Mealy machine.	172
9.7	A block diagram of the GCD calculator.	174
9.8	Elements of the ASM notation.	181
9.9	FSMD of the GCD calculator.	182
9.10	Communicating FSMs.	185

Chapter 10

10.1	Target Architecture.	193

List of Figures

10.2	A D-type flip-flop.	195
10.3	Type 1 Synchronous Reset (a) Synthesized Circuit (b) Timing Diagram.	197
10.4	Type 2 Synchronous Reset (a) Synthesized Circuit (b) Timing Diagram.	199
10.5	A D-type flip-flop with asynchronous reset.	202
10.6	A 4-bit register.	203
10.7	A gate level schematic for the simple sequential circuit.	206
10.8	Algorithmic description of an eight-bit GCD calculator.	207
10.9	Algorithmic Description of an eight-bit two's-complement multiplier.	208
10.10	Algorithmic description of an 8-bit (modified) Booth multiplier.	211
10.11	Target architecture (a) Process communication (b) FSMD architecture.	214
10.12	Two way handshaking communication.	216
10.13	Source oriented data communication.	218

Chapter 11

11.1	(a) A random access memory (b) Memory read timing specification (c) Memory write timing specification.	228
11.2	A diagram of controller, interface and memory module.	230
11.3	Timing of a memory interface.	232

Chapter 12

12.1	A mixed-level design description.	244
12.2	Synthesis from high level abstraction to low level.	245
12.3	Design representation for a single pulser (a) Timing diagram (b) Algorithmic level (c) State machine level (d) Register transfer level (e) Gate level.	247
12.4	The library compiler.	253
12.5	Delay calculation.	254
12.6	The MEBS system.	257
12.7	Differential equation solver (a) Control flow graph (b) Data flow graph.	260
12.8	Design space curve of the differential equation solver.	262
12.9	Control/Data flow graph of the GCD calculator.	264

12.10 Design space curve of the GCD calculator. 265

Chapter 13

13.1 Schematics for multiply with constants. 280
13.2 Timer implementation using (a) a conventional binary counter (b) an Linear Feedback Shift Register. 286

Chapter 14

14.1 Waveform for Bit Clock Generator. 296
14.2 Traffic Light Controller. 298
14.3 A Vending Machine Controller. 302

PREFACE

The purpose of this book is to introduce VHSIC Hardware Description Language (VHDL) and its use for synthesis. VHDL is a hardware description language which provides a means of specifying a digital system over different levels of abstraction. It supports behavior specification during the early stages of a design process and structural specification during the later implementation stages.

VHDL was originally introduced as a hardware description language that permitted the simulation of digital designs. It is now increasingly used for design specifications that are given as the input to synthesis tools which translate the specifications into netlists from which the physical systems can be built. One problem with this use of VHDL is that not all of its constructs are useful in synthesis. The specification of delay in signal assignments does not have a clear meaning in synthesis, where delays have already been determined by the implementation technology. VHDL has data-structures such as files and pointers, useful for simulation purposes but not for actual synthesis. As a result synthesis tools accept only subsets of VHDL. This book tries to cover the synthesis aspect of VHDL, while keeping the simulation-specifics to a minimum.

This book is suitable for working professionals as well as for graduate or undergraduate study. Readers can view this book as a way to get acquainted with VHDL and how it can be used in modeling of digital designs. It can be used as a teaching tool for hardware design and/or hardware description languages at the undergraduate or graduate level classes. A user of this book is assumed to have some experience in Pascal or C programming and some working knowledge of digital hardware design.

The book treats synthesis from the standpoint of the MEBS system. MEBS is a mixed-level entry synthesis system that takes a VHDL description, synthesizes it into structural level, and translates it into a format for FPGA or standard-cell implementation. The input design description contains multiple communicating processes in the following three levels of description: algorithmic level, finite state machine with datapath (FSMD) level, and register transfer

(RT) level. A designer can specify modules at different levels of abstraction in a single VHDL description. The synthesis system will perform the scheduling, allocation, and/or register transfer level synthesis on the design, depending on the level of abstraction of the module description.

Overview of the Book

This book is organized into two logical sections. The first section (Chapter 1 to Chapter 6) of the book introduces the features of VHDL; the second section (Chapter 7 to Chapter 14) discusses the VHDL modeling at various levels of abstraction, the concept of synthesis, and how to write efficient VHDL design descriptions.

The first chapter is an introductory chapter which gives a brief history of VHDL and how VHDL is used in the design world. The second chapter introduces the main VHDL constructs in a VHDL design description. Chapter 3 introduces some basic elements of VHDL including the data types, operators, and expressions. These elements will be used throughout the book. Chapter 4 discusses sequential statements that may appear in a process or a subprogram. Chapter 5 discusses concurrent statements which may appear within an architecture. Chapter 6 introduces the concept of subprograms and packages. Some readers might want to skip Chapter 3 at first and then return to it when their VHDL models evolve to require the inclusion of user-defined data types.

Chapters 7, 8, 9, and 10 introduce the modeling of a digital design at different levels of abstraction. Chapter 7 discusses how VHDL can be used to describe the structure of a hardware device. Chapters 8, 9, and 10 discuss how VHDL can be used to describe the behavior of hardware at RT level, FSMD level, and algorithmic level, respectively. Chapter 11 discusses about modeling and use of memories in VHDL. Chapter 12 discusses the process of translating a VHDL description downto to the gate level and also how to calculate the timing of a circuit. Chapter 13 discusses how an efficient VHDL design description can be developed. Finally, Chapter 14 goes through a few design examples as a way of illustrating the use of VHDL in digital design synthesis.

Acknowledgments

This book would not have been possible without the help of a number of people, and we would like to express our gratitude to all of them. We thank Professor James R. Armstrong at the Virginia Polytechnic Institute and State University, Professor Vijay K. Madisetti at the Georgia Institute of Technology, Professor Frank Vahid at the University of California, Riverside, and several anonymous reviewers who gave their careful review of the text. Their comments were both helpful and insightful.

The class notes used for CS/EE 120B Digital Systems formed the basis of the book. We would also like to thank Bassam Tabbara and Aruna Goli for proof reading the manuscript. We would also like to thank all the people who were involved in the MEBS project: ChengTsung Hwang who was involved in the early research on behavior synthesis, Tak Wang who developed the module generators for multiplication, Christy Yu who developed the graphical user interface, Shih-Hsu Huang who developed the early version of the scheduler, Chien Weng who developed the algorithms for subprogram synthesis, Stanly Ma who developed a printer interface card, and Aruna Goli who developed the design space exploration.

We would like to thank UC MICRO program, QuickTurn Design Systems, Inc., SMOS Systems, Inc., Fujitsu Laboratory of America, and the Xilinx, Inc. for their support. Special thanks to the following people who gave us valuable information and support during the development of the project. Ping Chao, Dr. Paul Lo, and Dr. T.C. Lin at the QuickTurn System, Inc., Dr. Chi-Ping Hsu at the Arcsys, Inc., Dr. Eugene Ko at National Semiconductor, Inc., Dr. Masahiro Fujita, Dr. K.C. Chen and Mike. T. Lee at the Fujitsu Laboratory of America, Junzo Tamada at the SMOS systems, Inc., and David Lam at the Xilinx, Inc.

Finally, we would like to thank Carl Harris at Kluwer Academic who made this book a reality.

VHDL MODELING FOR DIGITAL DESIGN SYNTHESIS

1
INTRODUCTION

One way to deal with the problem of the increasing complexity of electronic systems and the increasing time-to-market pressures is to design at high levels of abstraction. Traditional paper-and-pencil and capture-and-simulate methods have largely given way to the describe-and-synthesize approach for these reasons. In this chapter, we will first briefly overview the digital design process. Then, we will discuss the levels of abstraction of a design that have been used by designers. Hardware Description Languages (HDLs) have played an important role in the describe-and-synthesize design methodology. They provides a means of specifying a digital system at a wide range of levels of abstraction. They are used for specification, simulation, and synthesis of an electronic system. Among several different hardware description languages, VHSIC Hardware Description Language (VHDL) is one that is in widespread use today. It supports behavior specification at the early stages of a design process and structural specification at the later implementation stages. We will discuss the issues of modeling, simulation and synthesis of digital designs using VHDL.

1.1 DESIGN PROCESS

A design process of a digital system can be generally divided into the following four phases as illustrated in Fig. 1.1:

- Requirement analysis and specification phase
- Design phase
- Implementation and testing phase

- Manufacturing phase

The first phase in any digital system development project is the requirement analysis and specification phase, in which we determine and specify in detail the function, performance, and interface requirements of the system.

After the functionality of a digital system is specified, the next phase is the design phase. In a top-down approach to design, a digital system is usually partitioned into several less complex subsystems, which are in turn partitioned until, eventually, we end up with a detailed structural design which bounded to a specific technology. From top to down, a design phase may involve the following levels of decomposition: system design, architecture design, register transfer level (RTL) design. and logic level design. During the system design, a designer decomposes his/her system into several subsystems, and defines communication protocols among the subsystems. In an architecture design, the architectural style and performance of each subsystem is determined. During the RTL design phase, an architecture is translated into an interconnection of RTL modules. Finally, at the logic design phase, the RTL modules are constructed using logic gates.

During the implementation phase, the subsystems are implemented and tested. The implementation includes partitioning, placement, and routing to produce a layout of a circuit. These circuits are integrated into digital subsystems and tested. The subsystems are subsequently integrated into the entire digital system and a whole system testing is performed.

The final phase of a digital design process is to prototype, manufacture, and field test the design.

1.2 LEVELS OF ABSTRACTION

After we have clearly specified the problem at the global level, we seek a rational way to partition the problem into small pieces with clearly defined interrelationships. Our goal is to choose "natural' pieces in such a way that we can comprehend each piece as a unit and understand the interaction of the units. This partitioning process proceeds to lower levels until finally we choose actual gates. During the design decomposition process, it is important to maintain a proper level of view of the system – a level that provides the necessary in-

Introduction

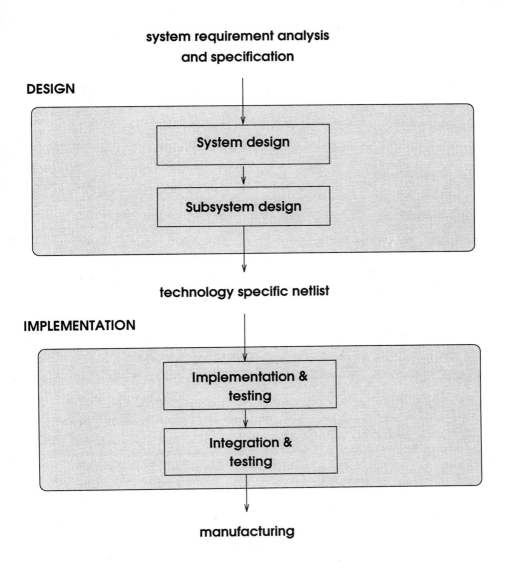

Figure 1.1 The Digital Design Process.

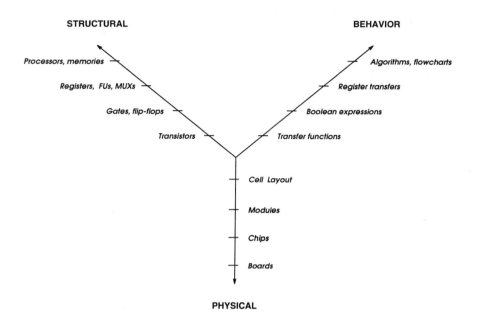

Figure 1.2 The Y-chart.

formation required at the moment, but does not overwhelm the design with unnecessary details.

In general, we may divide a design abstraction into the following levels: system level, register transfer level, logic level, and circuit level. To define and differentiate each abstraction level, we use the Y-chart (Fig. 1.2) to represent the abstraction of a design. The axes in the Y-chart represent three different domains of descriptions: behavior, structural and physical. In the behavior domain, we are interested in what a design does, not in how it is built; i.e. it describes the function of the design regardless of its implementation. In the structural domain, a design is described as an interconnection of components. A physical representation binds the structure to silicon. Along each axis the different levels of abstraction are described. As we move farther away from the center of the Y chart, the level of description becomes more abstract.

Depending on the design style and circuit complexity, designers can specify their designs in any or all of these design levels. Table 1.1 shows the primitive objects

Introduction

Levels of Abstraction	Behavior description	Structural description	Physical description
System level	Algorithms Flowcharts	Processors Memories	Boards Chips
RT level	Register transfers	Registers Function units Multiplexers	Chips Modules
Logic level	Boolean equations	Gates Flip-flops	Modules Cells
Circuit level	Transfer functions	Transistors Connections	Cells Wire segments

Table 1.1 Design objects on different levels of abstraction.

used on each level of abstraction. Models at different levels can be seen under different views. For example, at the system level, the primitive objects in the behavior domain are algorithms (processes) and communications (protocols); the primitive objects in the structural domain are an interconnection of processors, memories, and buses; while in the physical domain it is a layout of printed circuit boards or chips. At the register transfer level, a behavior description may be given as a set of state transitions and register transfers; a structural description views a design as an interconnection of registers, functional units and multiplexers; while a physical design may be a chip or a module. As we move down to lower levels of abstraction, the primitive objects become smaller and more detailed information of the objects will be involved.

1.3 DESIGN TOOLS

Fig. 1.3 shows the evolution of design tools during the past twenty years. Due to its good properties such as low power consumption and high packing density, CMOS technology has become a dominant technology for Application Specific Integrated Circuit (ASIC) design in late 1970s. Because a fabrication process is so expensive, the die size of a design is an important measure of economy. The goal of a designer is to minimize the area of a layout for a chip. A full custom layout methodology, where designers draw the layout of a module using layout editors, is the main design methodology used in this era. This methodology is

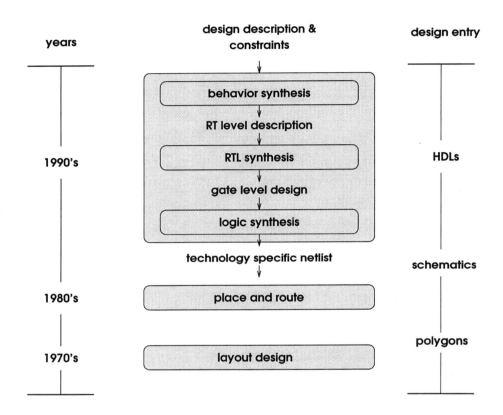

Figure 1.3 Design tools.

very time consuming and error prone because a design has to go all the way from functional specification down to layout design.

With the introduction of standard cells and gate arrays in the late 70's, ASIC designers turned to cell-based design methodologies for its design ease and its fast turn around time. Capture-and-simulate is the technique used for these layout methodologies. In this methodology, cells are pre-laid out, characterized, and put in the library. A designer manually translates his design from functional specification down to gate level, and uses a schematic editor to capture and simulate the gate-level design. After the simulation is completed, physical design can be done automatically by using placement and routing tools. Using this methodology, a design has to be translated manually from functional specification to the gate-level design which is still very time consuming .

The success in logic and sequential synthesis tools in late 1980s introduced the HDL-based design methodologies. Logic and sequential synthesis tools alleviate designers' effort in optimizing and mapping logic equations into gates. With these tools, designers describe their designs as a set of logic equations or a finite state machine in HDL, and use the synthesis tools to optimize it into structural netlists. In dealing with complex designs, this design methodology has the drawback that a large amount of detail is involved in writing the activities at the register transfer level.

As ASIC designs are getting more complicated and the request for shorter turn around time is getting stronger, designers are pushing towards higher levels of abstraction such as system level. Behavior synthesis is a process of converting a procedural-like algorithmic description into an RT level design. System synthesis is one which converts a specification into several communicating subsystems. To meet the problem of the increasing complexity of electronic systems and the increasing time-to-market pressures, there is a need to describe and synthesize a design at high levels of abstraction.

1.4 VHSIC HARDWARE DESCRIPTION LANGUAGES

HDL-based design methodologies have been getting popular and accepted by hardware designers in the past few years. Hardware description languages (HDLs) are mainly used by designers to describe the structure or behavior of a digital system for simulation or for synthesis. An analogy of using HDL for hardware design can be made to the history of using high-level programming languages for software design, where it starts from machine code (layered rectangles), to assembly languages (schematic), and to high-level languages (HDLs).

HDL-based design methodology has several advantages over traditional gate-level design methodologies. First, it improves the productivity. It allows a designer to create a design in less time and it allows people, without having much knowledge in circuit design, to design hardware. Second, it is portable to different technologies. HDL descriptions provide technology-independent documentation of the circuit. Using a synthesis tool, we can automatically convert an HDL description into several gate-level implementations for different technologies.

VHDL is one of few HDLs in widespread use today. It is recognized as a standard HDL by IEEE (IEEE standard 1076) and by the United States Department of Defense (MIL-STD-454L).

VHDL is a strongly typed language with a broad set of constructs. It supports a mixed-level description where structural or netlist constructs can be mixed with behavioral or algorithmic descriptions. With this mixed-level capability, we can describe a design where the existing modules are specified at the structure level (design reuse) and the newly developed modules are specified at the behavior level.

VHDL enables the description of a digital design at different levels of abstraction such as algorithmic, register transfer, and logic gate level. A user can abstract a design, or hide the implementation details of a design using VHDL. In a top-down design methodology, a designer usually represents a system in high level abstraction first and later converts it to a more detailed design. The breadth of VHDL allows one language to be used for the entire design process.

Fig. 1.4 shows how VHDL is used to describe a one-bit comparator from different domains. The circuit consists of two input ports A and B, and an output port C which is described in the entity (Fig. 1.4(a)). An entity defines a new component name, its input/output ports, and related declarations. An architecture specifies the relationships between the input ports and the outputs ports that may be expressed in term of behavior, data flow, or structure.

- A *behavioral style architecture* specifies what a particular system does in a program-like description, using VHDL processes, but provides no details as to how the design is to be implemented. For example, Fig. 1.4 (b) shown a behavior style architecture of a one-bit comparator.

- A *dataflow style* architecture specifies a system as a concurrent representation of the flow of control and movement of data. It models the information flow or dataflow behavior as combinational logic functions such as adders, comparators, decoders, and primitive logic gates. For example, Fig. 1.4 (c) shown a data flow style architecture of a one-bit comparator.

- A *structural style* architecture defines the structural implementation using component declarations and instantiations. Structural description includes a list of concurrently active components and their interconnections. For

Introduction

```
entity COMPARE is
  port ( A, B :   in BIT ;   C :   out BIT) ;
  end COMPARE ;
```

(a)

```
architecture BEHAVIOR  of COMPARE  is
begin
  process(A, B)
  begin
    if(A = B)  then C <= '1' ;    else C <= '0' ;
    end if ;
  end process;
end BEHAVIOR;
```

(b)

```
architecture DATAFLOW  of COMPARE  is
begin
   C <=  not(A  xor B)   after 10  ns;
end DATAFLOW ;
```

(c)

```
architecture STRUCTURE  of COMPARE  is
  component XOR_Gate
      port (I0, I1 :   in BIT ;  O :   out BIT) ;
  end component ;
  component NOT_Gate
      port (I0 :   in BIT ; O :   out BIT) ;
  end component ;
  signal NET_I : BIT ;
begin
  U0  :   XOR_Gate   port map(I0=>A, I1=>B, O=>NET_I) ;
  U1  :   NOT_Gate   port map(I0=>NET_I, O=>C) ;
  end STRUCTURE ;
```

(d)

Figure 1.4 Comparator (a) Entity (b) Behavior style (c) Data flow style (d) Structural style

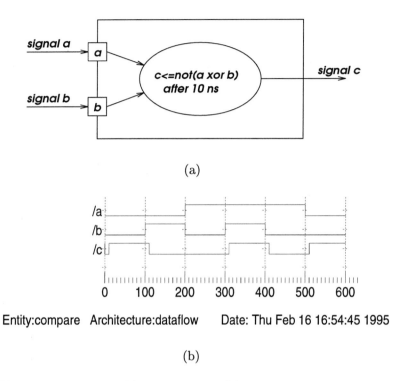

Figure 1.5 Simulation (a) hardware model (b) simulation waveform.

example, Fig. 1.4 (d) shows a structural style architecture of a one-bit comparator.

1.5 SIMULATION

A design description written in VHDL is usually run through a VHDL simulator to verify the behavior of the modeled system. To simulate a design, a designer must supply the simulator with a set of stimuli. The simulation program applies the stimuli to the input description at the specified times and generates the responses of the circuit. These results are interpreted by the designer to verify whether or not the design is satisfactory.

Introduction 11

A simulator can be used at any stage of a design process. At a higher level of the design process, simulation provides information regarding the functionality of the system under design. Normally, the simulation at this level is very quick but does not provide detailed information about the circuit functionality and timing. As a design process goes down to a lower level, the simulation will take longer time. Simulation at the lower level of a design process, runs much more slowly, but provides more detailed information about the timing and functionality of the circuit. VHDL allows mixed-level design in which some of the modules are described in a high-level of abstraction and some in a lower level of abstraction. The advantage of mixed-level simulation is that a designer can focus on the design of the timing critical modules, while leaving the non-critical modules to a later stage. To avoid the high cost of lower level simulation runs, simulators should be used to detect design flaws as early in the design process as possible.

During simulation of a VHDL program, a designer has to provide a set of functional test vectors associated with the different simulation time. Take the data flow model of the one-bit comparator as an example. Fig. 1.5 (a) shows its behavior model. The model responds to the events that occurs to either signal a or b. At simulation time 0, (a, b) is assigned with (0,0), the output (signal c) will change to 1 after 10 ns. At simulation time 100, signal b is changed to 1, hence the output will change to 0 after 10 ns of the current simulation time. The simulation continues and stops at simulation time 600. Fig. 1.5 (b) shows a simulation result.

1.6 SYNTHESIS

We define synthesis as a translation of a design description from one level of abstraction down to another level of abstraction. It could be a behavior to behavior translation, or behavior to structure. The translation process is similar to the compilation of high level programming language like C into assembly code. The inputs to a synthesis tool usually include an HDL description, timing and area goals, and a technology library. The outputs are the optimized netlists, estimated performance, and the area of the synthesized design. We will now briefly describe behavior synthesis, register transfer level(RTL) synthesis, and logic synthesis.

- Behavior synthesis is a process of translating a C-like algorithmic description into an RTL description. A design at the RT level usually includes (a)

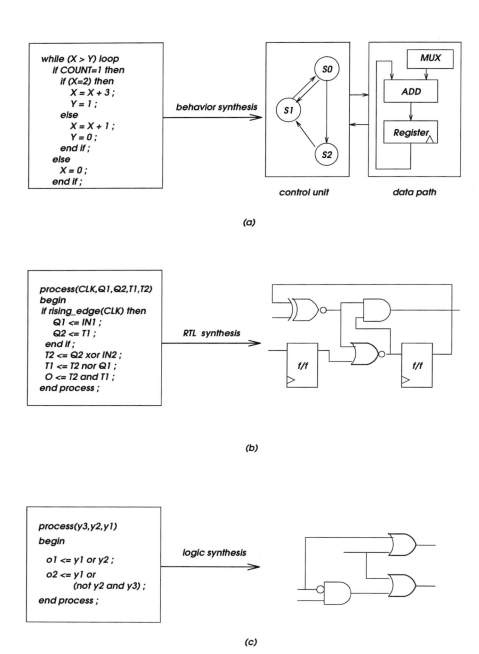

Figure 1.6 Synthesis (a) Behavior (b) RTL (c) Logic.

Introduction

datapaths (b) memories and (c) control units. Figure1.6 (a) illustrates a behavior synthesis process. This is also called *high-level synthesis* or *architectural synthesis*. The tasks include datapath synthesis, control synthesis, and memory synthesis.

- RTL synthesis is a process of generating structural netlists for a sequential circuit from a set of register transfer functions. The state boundary (clock by clock behavior) is already defined at this level. The register transfer operations can be described as a finite state machine or a set of register transfer level equations. The tasks includes state minimization, state encoding, logic minimization, and technology mapping.

- Logic synthesis translates Boolean expressions into combination logic. The optimization of logic synthesis is usually divided into two phases: a front end tool which minimizes the circuit, independent of the technology library, and a back end tool which maps a structural netlist to an interconnection of library cells.

1.7 SUMMARY

In this chapter, we discussed the four phases in the digital design process – requirement analysis and specification phase, design phase, implementation and testing phase, and manufacturing phase. We used the Y-chart to represent the various levels of abstraction and the different views of a design. We also discussed the modeling, simulation, and synthesis using VHSIC hardware description language. We discussed the trends in design automation of electronic systems, and concluded that there is a need to for a design synthesis at a high level of abstraction in order to deal with the problem of the increasing complexity of electronic systems and the increasing time-to-market pressures.

Exercises

1. Discuss the advantages and disadvantages of
    (a) capture-and-simulate method, and
    (b) describe-and-synthesize method.
2. Discuss the advantages and disadvantages of
    (a) behavior synthesis,
    (b) RTL synthesis,
    (c) logic synthesis.
3. Define the behavior synthesis, the RTL synthesis, the logic synthesis by drawing arcs in the Y-chart.
4. Explain the difference between behavior and structure of a digital design. Use a simple three input NAND gate to illustrate (a) a purely structural description, and (b) a purely behavior description.
5. Write a behavior description for a half adder.
6. Write a dataflow description for a half adder.
7. Write a gate level description for the half adder using an XOR gate and an AND gate.

2
BASIC STRUCTURES IN VHDL

In this chapter, we will look at the main VHDL constructs which are used in writing VHDL design descriptions. A digital system is usually designed as a hierarchical collection of modules. Each module corresponds to a design *entity* in VHDL. A design *entity* represents a portion of a hardware design that has well-defined inputs and outputs and performs a well-defined function. Each design entity has two parts: an entity *declaration* and *architecture* bodies. An entity *declaration* describes a component's external interface and *architecture* bodies describe its internal implementations. We can use *packages* in a design description. *Packages* define global information that can be used by several entities. A *configuration* binds component instances of a structure design into entity-architecture pairs. It allows a designer to experiment with different variations of a design by selecting different implementations. A VHDL design consists of several library units, each of which is compiled and saved in a design *library*.

This chapter defines the basic building blocks of a VHDL description:

- Entity
- Architecture
- Package
- Configuration
- Library

2.1 ENTITY DECLARATIONS

The entity declaration provides an "external" view of a component. It defines an interface to a component and describes the properties of the component's ports that can be seen from the outside, but does not provide information about how a component is implemented. The syntax [1] of an entity declaration is

>   **entity** *entity_name* **is**
>   [*generic_declaration*]
>   [*port_clause*]
>   { *entity_declarative_item* }
>   [**begin**
>   *entity_statement_part*]
>   **end** [*entity_name*];

where the *generic_declaration* declares constants that can be used to control the structure or behavior of the entity. The syntax is

>   **generic** (
>   *constant_name* : *subtype_indication* [:= init_value]
>   {; *constant_name* : *subtype_indication* [:= init_value] }
>   );

where *constant_name* specifies the name of a generic constant, *type* specifies the data type of the constant, and *init_value* specifies an initial value of the constant.

The *port_clause* specifies the interface channels of the entity, and its syntax is

>   **port** (
>   *port_name* : [*mode*] *subtype_indication* [:= *init_value*]
>   { ; *port_name* : [*mode*] *subtype_indication* [:= *init_value*] }
>   );

---

[1] In the notations for various syntax rules in this book, [ ] (square bracket) denotes optional parameters, | (vertical bar) indicates a choice among alternatives, and { } indicates that a choice of none, one or more items can be made.

where *port_name* specifies the name of a port, *mode* specifies the direction of a port signal, *subtype_indication* specifies the data type of a port or a generic constant and *init_value* gives an initial value to a port.

In an entity declaration, the entity itself and each of its ports is named with an identifier. VHDL identifiers are not case-sensitive. Some identifiers such as **entity, port, is**, and **end** are *reserved words* in VHDL. These identifiers have a fixed meaning in the language and may not be used for other purposes.

The ports are signals that communicate with the outside world. Each port is associated with a mode (**in, out, buffer, inout**) and a data type. The four port modes are

- **in:** can only be read. It is used for input only.
- **out:** can only be assigned a value. It is used for output only.
- **buffer:** can be read and assigned a value. It can have only one driver.
- **inout:** can be read and assigned a value. It can have more than one driver.

Signals with **buffer** property are normally used in the case when we want to read and write them from inside of a program, but only want to read them from outside of a program.

The *entity_declarative_item* declares some constants, types, or signals that can be used in the implementation of the entity, and the *entity_statement_part* contains the concurrent statements which are used for checking the operation conditions of a design entity.

Fig. 2.1(a) shows the interface of a one-bit adder. The entity name of the component is FULL_ADDER. It has input ports A, B, and CIN which are of data type BIT, and output ports SUM and COUT which are also of type BIT. A corresponding VHDL description is shown below.

```
entity FULL_ADDER is
   port( A, B, CIN  : in  BIT ;
         SUM, COUT : out BIT ) ;
end FULL_ADDER ;
```

We can control the structure and timing of an entity using generic constants. For example, in the following VHDL description generic constant N is used to

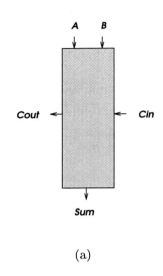

(a)

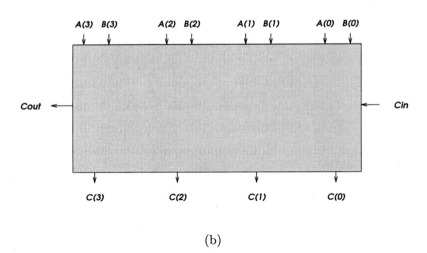

(b)

Figure 2.1 Entity declarations (a) one-bit adder (b) four-bit adder.

specify the number of bits for the adder and M is used to specify the timing behavior of the entity. N is defined as a generic constant with an initial value of 4. A corresponding interface for the four-bit adder is shown in Fig. 2.1(b). During the simulation or the synthesis process, the actual value for each generic constant can be changed depending on the actual needs of a design.

```
entity ADDER is
    generic(N: INTEGER := 4 ;
            M: TIME := 10 ns ) ;
    port( A, B, CIN : in  BIT_VECTOR (N-1 downto 0) ;
          CIN       : in  BIT ;
          SUM       : out BIT_VECTOR (N-1 downto 0) ;
          COUT      : out BIT ) ;
end ADDER ;
```

2.2 ARCHITECTURES

Once a module has had its interface specified in an entity declaration, one or more implementations of the entity can be described in *architecture* bodies. An architecture provides an "internal" view of an entity. It defines the relationships between the inputs and the outputs of a design entity which may be expressed in terms of behavior, dataflow, or structure.

An architecture determines the function of an entity. It consists of a declaration section where signals, types, constants, components, and subprograms are declared, followed by a collection of *concurrent statements*. The concurrent statements can be concurrent signal assignment statements, block statements and component instantiation statements. A concurrent statement communicates with other concurrent statements through signals. Each concurrent statement in an architecture defines a unit of computation that reads signals, perform computation based on their value, and assigns computed values to signals. They all together define interconnected pieces of hardware (in structural or behavioral forms) that jointly describe the overall structure and/or behavior of a design entity. The concurrent statements are executed in parallel. It does not matter on the order they appear in the architecture.

An architecture is declared using the following syntax:

architecture *architecture_name* **of** *entity_name* **is**

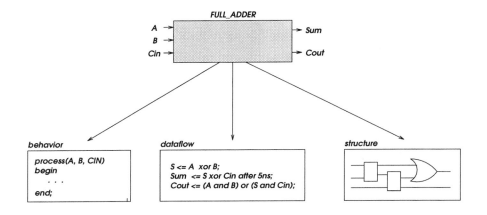

Figure 2.2 Entity and architectures.

$$\{ \text{ architecture\_declarative\_part } \}$$
begin
$$\{ \text{ concurrent\_statement } \}$$
end [*architecture_name*];

where the *entity_name* is the name of the entity being implemented, the *architecture_declarative_part* contains declarations to be used in the architecture definition, the *concurrent_statements* define pieces of hardware in structural or behavior forms. *Signals* are used to connect the separate pieces of an architecture. Each signal is associated with a *data type* that determines the kind of data it carries.

An entity may have several architectures. A designer may model a design using different implementation methods, at different views, or at different levels of abstraction. For example, for an adder design, one may describe two architectures: one uses a carry-look-ahead adder and the other uses a ripple-carry adder. A design may be described an adder in different views. One architecture migh be behavioral, while another might be a structural description. An architecture can also have a mixed design levels or views. Fig. 2.2 illustrates three alternate architectures of entity FULL_ADDER.

Basic Structures in VHDL

By convention, we can classify the styles of a description into behavior style, data flow style, and structure style.

2.2.1 Behavioral style

A *behavioral style* specifies what a particular system does in a program-like description using processes, but provides no details as to how a design is to be implemented. The primary unit of a behavior description in VHDL is the *process*. A process can be viewed as a program; it is constructed out of procedural statements and can call subprograms much as a program written in a general purpose procedural language like C.

The example below illustrates BEHAVIOR, which is an architecture of entity FULL_ADDER. It contains a process with a *sensitivity list* including signals A, B, and CIN. The execution of a process is suspended if there is no event occurs to the signals in the sensitivity list. Whenever there is a signal in the sensitivity list changes value (an event occurs on a signal), the process is reactivated and the statements inside the process will be executed sequentially.

```
architecture BEHAVIOR of FULL_ADDER is
begin
   process(A, B, CIN)
   begin
      if (A='0' and B='0' and CIN='0') then
         SUM  <= '0' ;
         COUT <= '0' ;
      elsif (A='0' and B='0' and CIN='1') or
            (A='0' and B='1' and CIN='0') or
            (A='1' and B='0' and CIN='1') then
         SUM  <= '1' ;
         COUT <= '0' ;
      elsif (A='0' and B='1' and CIN='1') or
            (A='1' and B='0' and CIN='1') or
            (A='1' and B='1' and CIN='0') then
         SUM  <= '0' ;
         COUT <= '1' ;
      elsif (A='1' and B='1' and CIN='1') then
         SUM  <= '1' ;
```

```
            COUT <= '1' ;
        end if ;
    end process ;
end BEHAVIOR ;
```

A process represents the behavior of some portion or all of a design. It defines an independent sequential body of code which can be activated in response to a change of state. When there are more than one process in an architecture, they will be executed concurrently. Inside a process, however, statements are interpreted sequentially. In the following, we will discuss a behavior model and a timing model of VHDL.

Behavior Model

A digital design is modeled as a collection of operations applied to values that are passed to the system. In a behavior model of VHDL, each operation is referred to as a *process* and the pathways through which values are passed to the system as *signals*. A system can be viewed as a set of processes, with all communications between processes taking place over these signals. All the processes in a model are said to be executing concurrently and signals are used to coordinate the concurrent processes.

A process can be viewed as an infinite loop. It starts with the first statement, then the second statement sequentially, until the last statement and goes back to the first statement. It continues to execute the statements in the process until it is suspended by **wait** statements. Once suspended, a process can be reactivated. One way to reactivate a process is by designating a maximum time for the process to remain suspended. Another way to reactivate a process is to wait until the system changes state and certain conditions are met. Such a change is reflected by a change in the value of a signal for a process since signals contain the state of the system. VHDL provides a means for a process to express its *sensitivity* to the value on the signals. A collection of all the sensitive signals of a process is called the *sensitivity list* of the process. Whenever a signal in the sensitivity list changes value, the process is reactivated.

Fig. 2.3(a) shows a representation of a system consisting of three processes in an architecture. Process i and process j each has two sensitive signals and process k has one sensitive signal. Process i and process j write to the same output signal. The three processes execute concurrently and the sensitive signals are used to coordinate the execution of the concurrent processes.

Basic Structures in VHDL 23

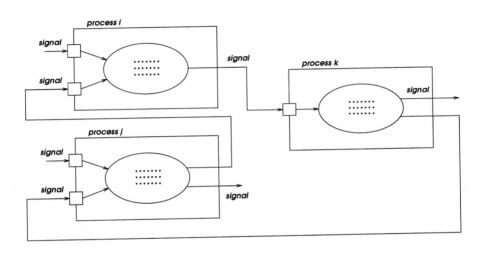

(a)

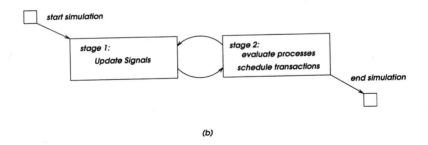

(b)

Figure 2.3 Behavior Model (a) Process model (b) simulation cycle.

Timing Model

Changes to signals occurred at given times are designated by the "simulation time" of the system. The simulation time at which something occurs to a signal in VHDL is different from the time of the platform's internal clock. A process reacts whenever there is a change to any signal in the sensitivity list. When a process generates a value on its output signals it may also designate the amount of time before the value is sent over to the output signal; this is referred to as *scheduling a transaction* after the given simulation time. It is possible to schedule any number of transactions for an output signal. The collection of all transactions for a signal in a process is called a *driver* of the signal. The driver is a set of time/value pairs which hold the value of each transaction and the time at which the transaction should occur.

VHDL has a two-stage model of time which is referred to as the *simulation cycle* as shown in Fig. 2.3(b). During the first stage of the simulation, the values of those signals which have transactions on the current simulation time are updated. During the second stage, those processes which receive information on their sensitivity list are evaluated (transactions are scheduled if any), until they are suspended. This stage is completed when all active processes are suspended. At the completion of the second stage, the simulation time is set to the next simulation time at which a transaction is to occur and the cycle is started again.

A designer may designate the amount of time before the value is sent over to the output signals in a signal assignment statement. If no delay is given in the assignment of a value to the signal, a *delta* delay is used to schedule a transaction. This delay does not update the time of the simulation clock but does require passing of a new simulation cycle. If the new value being assigned to a signal is different from the previous value of the signal, an *event* is said to occur to the signal.

2.2.2 Dataflow style architecture

A *dataflow style* architecture specifies a system as a concurrent representation of the flow of control and movement of data. It models the information flow or dataflow behavior, over time, of combinational logic functions such as adders, comparators, decoders, and primitive logic gates. The example below illustrates an architecture DATAFLOW of entity FULL_ADDER. In this architecture, there are three concurrent signal assignment statements. Each statement can

Basic Structures in VHDL

be treated as a process with the signals in the right hand side of the assignment as the sensitivity list of the process. For example, in the first assignment S gets the value of A **xor** B after an event occurs to either A or B. Here no time delay is specified, a delta delay is assumed by the simulator. In the second assignment, SUM gets the value of S **xor** CIN 10 ns after an event occurs to either S or CIN. Similarly, COUT gets value of ((A and B) or (S and CIN)) 5 ns after any event occurs to signal A, B, S or CIN.

```
architecture DATAFLOW of FULL_ADDER is
    signal S : BIT ;
begin
    S    <= A xor B ;
    SUM  <= S xor CIN after 10 ns ;
    COUT <= (A and B) or (S and CIN) after 5 ns ;
end DATAFLOW ;
```

Generic constants can be used as a delay parameter. The following example shows the declaration of a delay parameter of an entity declaration and its use in an architecture.

```
entity FULL_ADDER is
    generic(N: TIME := 5 ns) ;
    port( A, B, CIN  : in  BIT ;
          SUM, COUT  : out BIT ) ;
end FULL_ADDER ;
architecture DATAFLOW of FULL_ADDER is
    signal S : BIT ;
begin
    S    <= A xor B ;
    SUM  <= S xor CIN after 2*N ;
    COUT <= (A and B) or (S and CIN) after N ;
end DATAFLOW ;
```

2.2.3 Structural style architecture

A *structural style* architecture defines the structural implementation using component declarations and component instantiations. Structural description includes a list of concurrently active component instances and their interconnections. For example, the following shows a structural description of a one-bit full adder. Two types of components are defined in this example – HALF_ADDER and OR_GATE. The structural design of the full adder includes two HALF_ADDER instances and one OR_gate instance. These instances are connected by signals.

```
architecture STRUCTURE of FULL_ADDER is
   component HALF_ADDER
      port( L1, L2      : in  BIT ;
            CARRY, SUM  : out BIT ) ;
   end component ;
   component OR_GATE
      port( L1, L2 : in  BIT ;
            O      : out BIT ) ;
   end component ;
   signal N1, N2, N3 : BIT ;
begin
   HA1 :    HALF_ADDER  port map (A, B, N1, N2) ;
   HA2 :    HALF_ADDER  port map (N2, CIN, N3, SUM) ;
   OR1 :    OR_GATE     port map (N1, N3, COUT) ;
end STRUCTURE ;
```

As the complexity of a design increases, it becomes very difficult for a designer to represent a whole system as a flattened netlist. A designer usually decomposes his/her design into a set of functionally related subsystems. Each subsystem itself is a unit which can be further decomposed into a netlist of components. In VHDL, a top-level VHDL design which uses the structural style architecture consists of a set of instances connected by signals.

A component instance is a black box in the above description. Only the input and output connections are specified. A component instance needs to be bound to an entity which describes its functionality either in structural or behavioral description. Fig. 2.4(a) shows a hierarchy of entity-architecture pairs of a full adder design. The top-level entity consists of a netlist of two HALF_ADDER instances and a OR_GATE instance. The HALF_ADDER instance can be bound to another entity which consists of a netlist of an XOR gate and an AND gate.

```
entity HALF_ADDER is
   port( I0, I1 : in BIT ;
```

Basic Structures in VHDL

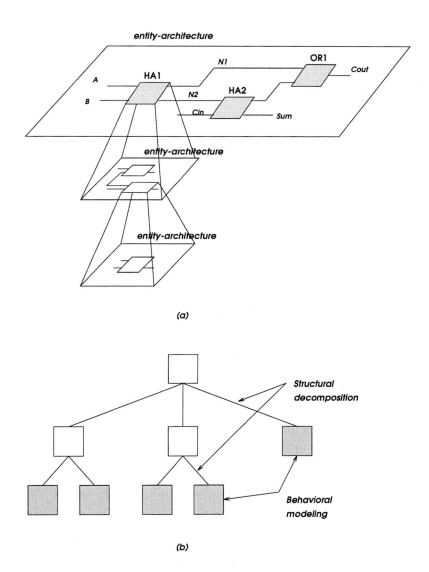

Figure 2.4 Structural Decomposition (a) A Design Hierarchy (b) Design Tree.

```
            S, Co  :   out BIT );
   end HALF_ADDER ;
   architecture STRUCTURE of HALF_ADDER is
      component XOR_GATE
         port (I0, I1 :  in BIT; O : out BIT ) ;
      end component ;
      component AND2_GATE
         port( I0, I1 :  in  BIT; O : out BIT ) ;
      end component ;
   begin
      U1 :    XOR_GATE    port map (I0, I1, S) ;
      U2 :    AND2_GATE   port map (I0, I1, Co) ;
   end STRUCTURE ;
```

Each of the component instances in the above description can be bound to another entity which describes its functionality. For example, the following entity describes the behavior of an XOR_GATE.

```
   entity XOR_GATE is
      port (I0, I1 :  in BIT ;
                 O : out BIT ) ;
   end XOR_gate ;
   architecture BHV of XOR_GATE is
   begin
      O <= I0 xor I1 after 10 ns ;
   end BHV ;
```

A structural form of the design hierarchy implies a design decomposition process. This is because at any level of the hierarchy, the system is composed of an interconnection of the component instances defined for that level. The structural description consists of a netlist of "black boxes". At the lowest level of the decomposition, the behavior of the the lowest-level design components must be specified. Fig. 2.4(b) shows the design tree of the full adder design. At the first level, a full adder is decomposed into two half adders and an OR gate. The half adders can be further decomposed into an exclusive-or gate and an AND gate. At the lowst level, a behavior description of each instance must be specified in order for the circuit description to be simulated.

2.3 PACKAGES

The primary purpose of a package is to collect elements that can be shared (globally) among two or more design units. A package may consist of two separate design units: a package declaration, and a package body. A package declaration declares all the names of items that will be seen by the design units that use the package. Typically, it contains some common data types, constants, and subprogram specifications.

A package body contains the hidden implementation details of the subprograms declared in the package declaration and the internal support of these subprograms. A package body is not required if no subprograms are declared in a package declaration.

The example below shows a package declaration. The package name is EX_PKG. It defines a data type INT8 which is an integer in the range of 0 to 255, two constants, ZERO and MAX, which are 0 and 100, respectively, and a function called Incrementer.

```
package EX_PKG is
    subtype   INT8 is INTEGER range 0 to 255 ;
    constant  ZERO  :  INT8 := 0 ;
    constant  MAX   :  INT8 := 100 ;
    procedure Incrementer (variable Count:  inout INT8) ;
end EX_PKG;
```

Because in package EX_PKG, we define a function called Incrementer, we need to define the behavior of the function separately in a package body. The separation between package declaration and package body serves the same purpose as the separation between entity declaration and architecture body.

```
package body EX_PKG is
    procedure Incrementer (variable Data :  inout INT8) is
    begin
       if (Count >= MAX) then
          Count := ZERO ;
       else
          Count := Count + 1 ;
       end if ;
    end Incrementer ;
```

end EX_PKG ;

2.4 CONFIGURATIONS

An entity may have several architectures. During the design process, a designer may want to experiment with different variations of a design by selecting different architectures. A configuration is a primary design unit which is used to bind component instances to different architectures. They can be used to provide fast substitution of component instances to entity binding of a structural design.

There are several ways of specifying the binding of a configuration. Chapter 7 has detailed description of each. The syntax of a configuration declaration is

> **configuration** *configuration_name* **of** *entity_name* **is**
>   { *configuration_declarative_part* }
> **for** *block_specification*
>   { *use_clause* }
>   { *configuration_item* }
> **end for** ;

The *configuration_declarative_part* of a configuration allows a configuration to use items from libraries and packages. The outmost *block_specification* in a configuration declaration defines the configuration for an architecture of the named entity. For example, the following shows a configuration of entity FULL_ADDER. The name of the configuration is FADD_CONFIG. The STRUCTURE refers to the architecture of entity FULL_ADDER to be configured. Under the architecture, instances HA1 and HA2 will be bound to entity HALF_ADDER architecture STRUCTURE in library WORK. Instance OR1 will be bound to entity OR_GATE in library WORK. Since there is no architecture specified for entity OR_GATE, the most recent compiled architecture of entity OR_GATE is assumed.

```
configuration FADD_CONFIG of FULL_ADDER is
  for STRUCTURE
    for HA1, HA2:   HALF_ADDER use entity WORK.HALF_ADDER(STRUCTURE)
    for OR1:   OR_GATE use entity WORK.OR_GATE ;
```

```
    end for ;
  end FADD_CONFIG;
```

2.5 DESIGN LIBRARIES

VHDL analysis is the process of checking a VHDL design description for syntactic and semantic correctness. After VHDL analysis, the design units of a VHDL file are kept inside a *design library* for subsequent use (normally for synthesis or simulation). A design library can contain the following *library units*:

- Packages - shared declarations
- Entities - shared designs
- Architectures - shared design implementations
- Configurations - shared design versions.

A library unit is a VHDL construction that can be separately analyzed. Note that "separately analyzable" does not mean "analyzable in any order". For example, an entity must be analyzed before its architectures, and a package must be analyzed before it is used by a design unit.

VHDL defines a special design library called "WORK". When you compile a VHDL program without specifying the target library, the VHDL program will be compiled into library WORK. For example, if the command for VHDL analysis is **vc**, then the following command

```
vc My_Design.vhd
```

will check the syntactic correctness of file "My_Design.vhd" and places the library units of the file into library **WORK**.

VHDL allows a designer to maintain multiple design libraries as shown in Fig. 2.5. To specify a specific design library to keep the library units of a file, use **-w** or **-work** option. For example, the following command compiles My_Design.vhd into library Library_1.

```
vc -w Library_1 My_Design.vhd
```

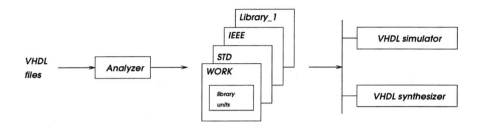

Figure 2.5 Design Libraries.

The names of a design library and the library units inside a design library are not automatically visible to other library units. A *library_clause* will make a library visible. The syntax of the *library_clause* is

   library *library_name* { , *library_name* } ;

where *library_name* is the name of a design library.

After a library is made visible, we need to make the declarations of the package we want to use visible. VHDL provides a *use_clause* to extend the visibility of those declarations to other design units. For example, if we have compiled package *My_PKG* into library *Library_1*. To use the items declared in the package, we can use the **use** statement to include part or all of the package.

```
library Library_1 ;
use Library_1.My_PKG.all ;
```

In order for a VHDL description to be installation-independent and/or portable, VHDL uses the concept of *logical library* and *physical library*. Inside a VHDL program, we use logical libraries, such as IEEE, STD, or WORK (default). During the simulation or synthesis process, a command is provided to indicate the exact location of the design library. For example, the following command indicates the logical library IEEE will be mapped to physical location /usr/cad/ieee.

```
logic_name   IEEE    /usr/cad/ieee
```

The above statements are usually the first in a package or entity specification source file. Note here, the effect of *library_clause* and *use_clause* are only applied to the adjacent design unit.

2.6 SUMMARY

1. A design entity has two parts: the declaration and the architecture. An entity declaration contains an interface to a design, which is implementation independent. An architecture defines the function of a design entity. It is always related to an entity.

2. Packages declare constants, data types, components, and subprograms that are shared by several designs and/or entities.

3. Configuration is used to bind a component instance of a structured design to an entity specification. It allows a designer to experiment with different versions of a design by changing different architectures.

4. *Packages, entities, architectures,* and *configurations,* are called library units in VHDL.

5. A VHDL description is compiled into an intermediate form and placed in a design library. The default library is called WORK.

Exercises

1. What are the reasons to separate entity declaration and architecture definition?
2. Can you nest entity and architecture declarations in your source design file?
3. Write an entity description for a one-bit D flip-flop.
4. Write an entity description for a N-bit shift register using generic constants.
5. Redesign the full-adder circuit using only two-input NAND gates. Write a dataflow description for your design. Assume each gate takes 20 ns. Use a generic constant in your design to control the delay of a NAND gate.
6. Write a gate level description for the full adder in the previous problem. Write a configuration which binds the NAND_2 gate instances to an entity (with entity name N2 and architecture IMP) in library B1.
7. What are the reasons to separate package declaration and package body?
8. Write an entity description for a BCD to seven-segment display converter. In the implementation, you need a decoder and a converter. Assume they are described as separate entities. Write a structural style architecture for the design.
9. When a design unit is compiled, where will it be stored?
10. What is a design library used for? Suppose you have a VHDL design description called **foo.vhd**. What is the meaning of **vc foo.vhd**? What is the meaning of **vc -w blib foo.vhd**?

3

TYPES, OPERATORS AND EXPRESSIONS

A data object holds a value of specific type. VHDL provides three classes of objects: variables, signals and constants. Declarations list the data objects to be used, state what type they are, and perhaps what their initial values are. An expression is a formula that defines the computation on the data objects. Expressions perform arithmetic or logical computations by applying an *operator* to one or more *operands*. *Operators* specify the computation to be performed, while *operands* are the data for the computation.

3.1 DATA OBJECTS

There are three classes of data objects in VHDL: constants, variables and signals. The class of an object is specified by a reserved word that appears at the beginning of the declaration of that object.

1. A constant is an object which is initialized to a specific value when it is created, and which cannot be subsequently modified. Constant declarations are allowed in packages, entities, architectures, subprograms, blocks and processes. The syntax for a constant declaration is:

    constant *constant_name* {, *constant_name* } : *type_name* [:= *value*] ;

    ```
    constant YES    : BOOLEAN := TRUE ;
    constant CHAR7  : BIT_VECTOR(4 downto 0) := "00111" ;
    constant MSB    : INTEGER := 5 ;
    ```

2. Variables are used to hold temporary data. They can only be declared in a *process* or a *subprogram*. A variable must declare a type and can be given

a range constraint or an initial value. By default, the initial value is the lowest (leftmost) value of range for that type. The syntax for a variable declaration is as follows:

variable *variable_name* {, *variable_name* } : *type_name* [:= *value*];

```
variable   STAND,BROKE  : BIT ;
variable   TEMP         : BIT_VECTOR(8 downto 0) ;
variable   DELAY        : INTEGER range 0 to 15 := 0 ;
```

3. Signal declarations create new signals (wires) of a given type. They can be used to communicate between processes or to synchronize processes. Signals can be declared in package declarations (global signals), entities (entity global signals), architectures (architecture global signals), and blocks. They can be used but cannot be defined in processes and subprograms. The syntax for a signal declaration is as follows:

signal *signal_name* {, *signal_name* } : *type_name* [:= *value*] ;

```
signal   BEEP     : BIT := '0' ;
signal   RESULT   : INTEGER range 0 to 1023 ;
signal   COUNT    : STD_LOGIC_VECTOR(7 downto 0) ;
```

3.2 DATA TYPES

All the data objects in VHDL must be defined with a data type. VHDL contains a wide range of types that can be used to create simple or complex objects.

A type declaration defines the name of the type and the range of the type. Type declarations are allowed in package declaration sections, entity declaration sections, architecture declaration sections, subprogram declaration sections, and process declaration sections. Data types include:

- Enumeration types
- Integer types
- Predefined VHDL data types
- Array types
- Record types

Types, Operators and Expressions 37

- STD_LOGIC data type
- SIGNED and UNSIGNED data types
- Subtypes

3.2.1 Enumeration Types

An enumeration type is defined by listing all possible values of that type. All of the values of an enumeration type are user-defined. These values can be identifiers or single-character literals. An identifier is like a name such as *white*, *monday* and *lily*. Character literals are single characters enclosed in quotes such as 'X', '0', and '1'. The syntax is

> **type** *type_name* **is** (*enumeration_literal* {, *enumeration_literal* }) ;

where *type_name* is an *identifier* and each *enumeration_literal* is either an identifier or a character literal. An enumeration literal can be defined in two or more enumeration types.

The following shows enumeration type definitions.

```
type COLOR is (RED, ORANGE, YELLOW, GREEN, BLUE, PURPLE) ;
type STD_ULOGIC is ('U','X','0','1','Z','W','L','H','-') ;
variable X : COLOR ;
signal   Y : STD_ULOGIC ;
```

Each identifier in a type has a specific position in the type determined by the order in which the identifier appears in the type. By default, the first identifier will have a position number of 0, the next a position number of 1, and so on. In hardware synthesis, the size of a signal or variable of an enumeration type is determined by the minimum number of bits required to encode the number of enumerated values.

3.2.2 Integer Types

Integer types (or subtypes) are for mathematical integers. All of the normal predefined mathematical functions like add, subtract, multiply, and divide apply to integer types. The syntax is

type *type_name* **is range** *integer_range* ;

where *type_name* is the name of the integer type, and *integer_range* is a subrange of the integer type. The following shows some integer declarations.

```
type INTEGER is range -2147483647 to 2147483647 ;
type TWOS_COMPLEMENT_INTEGER is range -32768 to 32767 ;
```

3.2.3 Predefined VHDL Data Types

IEEE 1076-1987 predefined two site-specific packages: the Standard and the Textio package, in the STD library. Each contains a standard set of types and operations. The Standard package is included in all VHDL source files by an implicit use clause. The following shows a summary of the data types defined in the Standard package.

- BOOLEAN : The BOOLEAN data type is an enumeration type with two values, **false** and **true**, where **false** < **true**. Logical operations and relational operations return BOOLEAN values.

- BIT : The BIT data type is an enumeration type with two values, '0' and '1'. Logical operations can take and return BIT values.

- CHARACTER : The CHARACTER data type is an enumeration type of ASCII character set. Nonprinting characters are represented by a three-letter name. Printable characters are represented by themselves in single quotation marks.

- INTEGER : The INTEGER data type represents positive and negative numbers. The range specified in the Standard package is from -2,147,483,647 to +2,147,483,647. All of the normal predefined mathematical functions like add, subtract, multiply, and divide apply to integer types.

- NATURAL : The NATURAL data type is a subtype of INTEGER, used for representing natural (non-negative) numbers.

Types, Operators and Expressions   39

- POSITIVE : The POSITIVE data type is a subtype of INTEGER, used for representing positive (nonzero, non-negative) numbers.

- BIT_VECTOR : The BIT_VECTOR data type represents an array of BIT values.

- STRING : The STRING data type is an array of CHARACTERs. A STRING value is enclosed in double quotation marks.

- REAL : Real types are used to declare objects that emulate mathematical real numbers. The range specified in the Standard package is from -1.0E+38 to +1.0E+38.

- Physical type TIME : The TIME data type represents a time value used for simulation.

3.2.4 Array Types

Array types group one or more elements of the same type together as a single object. Array elements can be of any VHDL data type. The number of indices (the dimension) of an array in VHDL be be any positive number. An array has one and only one *index* whose value selects each element. The range of the index (or called index range) determines the number of elements in the array and their direction (low to high or high downto low). In VHDL, an index can be of either an integer type or enumeration type.

Array data type can be divided into constrained array type and unconstrained array type. A constrained array type is a type whose index range is explicitly defined. The syntax of a constrained array type definition is

type *array_type_name* **is array** (*discrete_range*) **of** *subtype_indication*;

where *array_type_name* is the name of the constrained array type, *discrete_range* is a subrange of another integer type or an enumeration type, and *subtype_indication* is the type of each array element.

For example, the following defines an array type AR1 of 64 elements in which each element is of type INTEGER.

```
type AR1 is array (0 to 63) of INTEGER ;
```

An unconstrained array type is a type whose dimension is given, but the exact range and direction of each dimension is left unspecified. This allows multiple subtypes to share a common base type. The syntax of a unconstrained array type definition is

type *array_type_name* **is array** (*type_name* **range** <>) **of** *subtype_indication* ;

type_name is the subtype definition of the index.

For example, there are two predefined unconstrained arrya type in VHDL: BIT_VECTOR and STRING types. Their definitions are shown as follows.

```
type BIT_VECTOR is array (NATURAL range <>) of BIT ;
type STRING is array (POSITIVE range <>) of CHARACTER ;
```

To use an unconstrained array type, the index range has to be specified. For example

```
subtype B4 is BIT_VECTOR (3 downto 0) ;
variable V5 :  BIT_VECTOR(4 downto 0) ;
```

Arrays are used for modeling linear structures such as registers, RAMs and ROMs. The elements of an array can be addressed using an index variable. The index of an array must be a discrete type (integer type or enumeration type). The example below shows a definition of a 16×10 memory type.

```
subtype   INT4      is INTEGER range 0 to 15 ;
subtype   INT10     is INTEGER range 0 to 1023 ;
type      Memory    is array ( INT4 ) of INT10 ;
```

Each element of an array can be accessed by an array index. An example of how to access elements of the array is as follows.

```
variable X: Memory ;
variable I: INT4 ;
variable Y: INT10 ;
Y := X(I+3) ;
```

VHDL allows declarations of multiple dimensional arrays. The example below shows a declaration of a two-dimensional array type TWO_D_AR. In the assignment, a signle element '1' will be assigned to variable X.

Types, Operators and Expressions 41

```
type TWO_D_AR is array(0 to 7, 0 to 3) of BIT ;
constant TWO_D_ROM : TWO_D_AR :=
( ( '0', '0', '0', '1'),
  ( '0', '0', '1', '1'),
  ( '0', '1', '0', '1'),
  ( '0', '1', '0', '1'),
  ( '0', '0', '1', '1'),
  ( '1', '0', '0', '1'),
  ( '1', '1', '0', '1'),
  ( '1', '1', '1', '1') ) ;
X := TOW_D_ROM(2, 3) ;
```

3.2.5 Record Types

Record types group one or more elements of different types together as a single object. Each element of the record can be accessed by its field name. Record elements can include elements of any type, including array and records. An example of a record type declaration is as follows:

```
type DATE_TYPE is (SUN, MON, TUE, WED, THR, FRI, SAT) ;
type HOLIDAY is
  record
    YEAR: INTEGER range 1900 to 1999 ;
    MONTH: INTEGER range 1 to 12 ;
    DAY: INTEGER range 1 to 31 ;
    DATE: DATE_TYPE ;
  end record ;
```

A selected name is used to reference an element of a record. A selected name consists of the name of the object, followed by a dot (.), followed by the name of the element. The type of a selected name is the type of its element. For example,

```
signal S: HOLIDAY ;
variable T1:  INTEGER range 1900 to 1999 ;
variable T2:  DATE_TYPE ;
T1 := S.YEAR ;
```

```
T2 := S.DATA ;
```

3.2.6 STD_LOGIC Data Types

To model a signal line which may be presented by more than two values, VHDL defines a Standard Logic package (IEEE Std 1164-1993). Two of the basic data types in this package are STD_ULOGIC and STD_LOGIC. The STD_ULOGIC defines a data type containing nine values. This data type is unresolved. In other words, if a signal is declared as STD_ULOGIC type, a resolution function must be provided if it will be driven by multiple drivers. The STD_LOGIC data type is a resolved data type. A resolution function for this type has been provided in the Standard Logic package.

The nine values include:

```
type STD_ULOGIC is (
    'U'    --    Uninitialized
    'X'    --    Forcing Unknown
    '0'    --    Forcing Low
    '1'    --    Forcing High
    'Z'    --    High Impedance
    'W'    --    Weak Unknown
    'L'    --    Weak Low
    'H'    --    Weak High
    '-'    --    Don't Care
) ;
```

Similar to BIT and BIT_VECTOR types, VHDL provides STD_ULOGIC_VECTOR and STD_LOGIC_VECTOR. For example, the interface for a four bit adder can be described as

```
entity ADDER is
   port( A, B : in  STD_LOGIC_VECTOR (3 downto 0) ;
         CIN  : in  STD_LOGIC ;
         SUM  : out STD_LOGIC_VECTOR (3 downto 0) ;
         COUT : out STD_LOGIC ) ;
end ADDER ;
```

Types, Operators and Expressions 43

To use the definitions and functions of the Standard Logic package, the following statements have to be included in the program.

```
library IEEE ;
use IEEE.STD_LOGIC_1164.all ;
```

3.2.7 SIGNED and UNSIGNED Data Types

The SIGNED and UNSIGNED data types are used for objects which will be accessed by bits (such as BIT_VECTOR type) while at the same time can be manipulated as integers. Both the SIGNED and UNSIGNED data types are defined in the Standard Synthesis packages: NUMERIC_BIT and NUMERIC_STD. The NUMERIC_BIT package is based on type BIT, while the NUMERIC_STD package is based on type STD_LOGIC. The definitions of SIGNED and UNSIGNED data types in these two packages are

```
type SIGNED   is array (NATURAL range <>) of BIT/STD_LOGIC ;
type UNSIGNED is array (NATURAL range <>) of BIT/STD_LOGIC ;
```

To make the two data types available for arithmetic computation, overloading functions are provided in both packages. Objects with UNSIGNED type are interpreted as unsigned binary integers and objects with SIGNED type are interpreted as two's complement binary integers. In the following example, Card_Input is treated as a 4-bit unsigned number, and Score is treated as a 5-bit signed number.

```
Card_Input :    UNSIGNED(3 downto 0);
Score      :    SIGNED(4 downto 0) := "00100";
```

Both packages are compiled into the IEEE library. Since SIGNED and UNSIGNED are defined in both numeric packages, only one package can be used at a time or there will be a conflict. To use the definitions and functions of the NUMERIC_BIT package, the following statements have to be included in a program:

```
library IEEE ;
use IEEE.NUMERIC_BIT.all ;
```

To use the definitions and functions of the NUMERIC_STD packages, the following statements have to be included in a program:

```
library IEEE ;
```

```
use IEEE.STD_LOGIC.all ;
use IEEE.NUMERIC_STD.all ;
```

3.2.8 Subtypes

VHDL provides *subtypes* , which are defined as subsets of other types. Anywhere a type definition can appear, a subtype definition can also appear. The difference between a type and a subtype is that a subtype is a subset of a previously defined *base* type or subtype. Overlapping subtypes of a given base type can be compared and assigned to each other.

Subtypes are a powerful way to ensure valid assignments and meaningful data handling in VHDL. Subtypes inherit all operators and subprograms defined for their parent types.

For example, NATURAL and POSITIVE are subtypes of INTEGER and they can be used with any INTEGER function. They can be added, multiplied, compared and assigned to each other, as long as the values are within the appropriate subtypes's range. The example below shows some subtype declarations.

```
subtype INT4   is INTEGER range 0 to 15 ;
subtype BIT_VECTOR6  is BIT_VECTOR(5 downto 0) ;
```

3.3 OPERATORS

VHDL provides six classes of operators. Each operator in VHDL has a precedence level. Table 3.1 contains a concise list of the classes of the operators in order, from the lowest to the highest precedence.

All the operators in the same class have the same precedence level and are left-associative. Both the precedence and the associativity of an operator are fixed and there is no way to change them, but the parentheses can be used to control the association in an expression. Where parentheses do not explicitly indicate the grouping of operators with operands, the operands are grouped with the operator having higher precedence. If two operators are in the same class (have the same precedence), the operators are associated with their textual order, from left to right.

Types, Operators and Expressions 45

logic_operator	and, or, nand, nor, xor
relational_operator	=, /=, <, <=, >, >=
adding_operator	+, -, &
sign	+, -
multiplying_operator	*, /, mod, rem
miscellaneous_operator	**, abs, not

Table 3.1 Predefined Operators

3.3.1 Logical Operators

The logical operators **and, or, nand, nor, xor,** and **not** accept operands of pre-defined type BIT, type BOOLEAN, and one-dimensional array type of BIT. For the binary operators **and, or, nand, nor,** and **xor**, the base type of both operands must be the same, and if the type of the operands is a one-dimensional array, the operands must be arrays of the same length. The binary operators designate the bitwise function on matching elements of the arrays, and the result is an array with the same index range as the left operand. The unary operator **not** computes the bitwise negation (not) of its operand, and the result is an array with the same index range as the operand if the type of the operand is an array. The example below shows some logical signal declarations and logical operations.

```
signal  A, B, C         : BIT_VECTOR(6 downto 0) ;
signal  D, E, F, G, H   : BIT ;
A <= B and C ;
D <= (E xor F) and (G xor H) ;
```

3.3.2 Relational Operators

The binary operators "=", "/=", "<", "<=", ">", and ">=" are used to compare their operands. The two operands must be of the same type, and the result is a BOOLEAN value.

The "=" operator tests for the relationship "is equal to"; the "/ =" operator tests for the relationship "is not equal to"; the "<" operator tests for the relationship "is less than"; the "<=" operator tests for the relationship "is less than or equal to"; the ">" operator tests for the relationship "is greater than"; and the ">=" operator tests for the relationship "is greater than or equal to." The result is TRUE if the stated relationship holds for the particular operand values and FALSE if the stated relationship does not hold.

The equality ("=") and inequality ("/ =") operators are defined for all types. The equality and inequality operators are the inversion functions of each other. The ordering ("<", "<=", ">", ">=") operators are defined for all enumerated types, integer types, and one-dimensional arrays of enumeration or integer types. The internal order of a type's values determines the result of the ordering operators. Integer values are ordered from negative infinity to positive infinity. The relative order of two array values is determined by comparing each pair of elements in turn, beginning from the left bound of each array's index range.

In the following example, two operands of the same type are compared and the result of a comparison will be of BOOLEAN type.

```
signal A, B     : UNSIGNED(6 DOWNTO 0) ;
signal X, Y     : COLOR1 ;
signal C, D, E  : BOOLEAN ;
C <= (A = B) ;
D <= (X < Y) ;
E <= (C > D) ;
```

3.3.3 Adding Operators

Adding operators include adding "+", "−" and "&". The arithmetic operators ("+" and "−") are predefined for INTEGER operands. The concatenation operator "&" is supported for all register-array objects. The concatenation operator builds a register-array by combining the operands. Each operand of "&" can be an array or an element of an array. These operators also support data types SIGNED and UNSIGNED in the Standard Synthesis packages.

An unsigned number can operate with an integer. It can also operate with a bit vector. For example, In the following example, W is a BIT_VECTOR, X is UNSIGNED, and Y and Z are UNSIGNED numbers.

```
signal W    : BIT_VECTOR(3 downto 0) ;
signal X    : INTEGER range 0 to 15 ;
```

Types, Operators and Expressions 47

```
signal Y, Z : UNSIGNED(3 downto 0) ;
Z <= X + Y + Z ;
Y <= Z(2 DOWNTO 0) & W(1) ;
```

3.3.4 Sign Operators

The unary plus operator ("+") simply yields the value of its operand. The unary minus operator ("−") computes the arithmetic negation of its operand. The unary operator "**abs**" returns the absolute value of its operand. All three operators take an object of numeric type and return the same type.

3.3.5 Multiplying Operators

VHDL predefines the multiplying operators (including "******", "*****", "**/**", **mod**, and **rem**) for all INTEGER types. "******" represents exponentiation, "*****" represents multiplication, "**/**" represents division, **mod** represents modulus, and **rem** represents remainder. Except for "******" operation, these operators are also overloaded to support data types SIGNED and UNSIGNED in the Standard Synthesis packages.

The following examples show some statements using multiplying operators.

```
signal A, B, C, D, E : INTEGER range 0 to 31 ;
A <= B * B ;
C <= D / 8 ;
D <= E MOD 4 ;
F <= 100 REM 9 ;
```

3.4 OPERANDS

In an expression, the operator uses the operands to compute its value. There are many categories of operands. Operands can themselves be expressions.

The operands in an expression include:

- Literals : 'x', "1001", 345

- Identifiers : Red, stand
- Indexed names : Data_Array(8)
- Slice names : bitvec(8 downto 2)
- Attributes names : NATURAL'left
- Aggregates : (1, 2, 3), (1—3 => '1', others => '0')
- Qualified expressions : COLOR1'(white)
- Function calls : foo(i, j, k)
- Type conversions : BIT_VECTOR(var)

3.4.1 Literals

A literal (constant) is either a *numeric literal*, *a character literal*, *an enumeration literal*, or *a string literal*.

Numeric literals are constant integer values. The two kinds of numeric literals are *decimal* and *based*. A decimal literal is written in base 10. A based literal can be written in any base from 2 to 16 and is comprised of the base number, a (#), the value in the given base, and another (#); for example, 4#0_123# is decimal 27. The numeric literal can be separated by underscores.

Character literals are single characters enclosed in single quotation marks, for example: 'Z'.

Enumeration literals are literals defined in the enumeration type definition. They can be identifiers and character literals as shown below. The enumeration literal can be overloaded in VHDL. In other words, the same enumeration literal can belong to two or more enumeration types.

```
type PAINT_COLOR is (RED, YELLOW, BLUE) ;
type LIGHT_COLOR is (RED, YELLOW, GREEN) ;
```

String literals are representations of one dimensional arrays of characters. The two kinds of strings are *character strings* and *bit strings*. Character strings are a sequence of characters in double quotes such as "DEMO", and "10101". *Bit strings* are similar to character strings, but represent binary, octal, or hexadecimal values such as B"11001", O"277", and X"4C".

Types, Operators and Expressions 49

For example, 987 is a numeric literal, **BLUE** is an enumeration identifier, 'A' is a character literal, "DEMO" is a character string, and B"01001" is a binary bit string.

3.4.2 Identifiers

An identifier is sometimes called a simple name. An identifier is the name for a constant, a variable, a signal, an entity, a port, a subprogram, and a parameter declaration. Reserved words in VHDL are also identifiers. A name must begin with an alphabetic letter (a - z) followed by letters, underscores, or digits. Underscore ('_') cannot be the last character of an identifier. VHDL identifiers are *not case-sensitive*. This means VHDL does not distinguish x from X. Some identifiers such as **entity, port, is** and **end** are *reserved words* in VHDL. Each reserved word has a fixed meaning in the language and may not be used for other purposes in VHDL.

3.4.3 Indexed Names

An indexed name identifies one element of an array object. The syntax of an indexed name is

array_name (expression)

where *array_name* is the name of a constant or variable of an array type. The *expression* must return a value within the array's index range. The value returned to an operator is the specified array element.

The example below shows a declaration of an 8×10 memory. We use indexed name to access the elements of the memory.

```
type Memory is array (0 to 7) of INTEGER range 0 to 1023 ;
variable Data_Array :  Memory ;
variable ADDR       : INTEGER range 0 to 7 ;
variable DATA       : INTEGER range 0 to 1023 ;
variable X          : BIT_VECTOR (6 downto 0) ;
variable Z          : BIT ;
Z    := X(3) ;
DATA := Data_Array(ADDR) ;
```

3.4.4 Slice Names and Aliases

Slice names identify a sequence of elements of an array object. The direction can be either **to** or **downto**. However the direction of a slice must be consistent with the direction of the identifier's array type. A slice name can be used in combination with an alias declaration.

An *alias* creates a new name for all or part of the range of an array object. It gives an alternative way to access the objects. It is very useful for naming parts of a range as if they were subfields. Aliases provide a mechanism to name each of the subfields of the instruction and to reference these fields directly by the alias name. In the following example, ORG is declared as a bit vector. We can use **alias** to rename partial or the whole vector.

```
variable ORG : BIT_VECTOR(7 downto 0) ;
alias    AL1 :  BIT_VECTOR(0 to 3) is ORG(7 downto 4) ;
--   reverse direction
alias    AL2 :  BIT_VECTOR(107 downto 100) is ORG ;
--   change index range
alias    AL3 :  BIT is ORG(7) ;
--   single element alias
ORG := AL1(0 to 2) & AL2(106 downto 103) & AL3 ;
--   equal to ORG(7 downto 5) & ORG(6 downto 3) & ORG(7)
```

3.4.5 Attributes Names

An attribute is data that is attached to VHDL object.s A VHDL attribute takes a variable or signal of a given type and returns a value. The syntax of an attribute name is

perfix'attribute

The following are some commonly used predefined attributes: *left*, *right*, *low*, *high*, *range*, *reverse_range* and *length*. The *left* (*right*) attribute returns the index of the leftmost (rightmost) element of the data type. The *high* (*low*) attribute returns the index of the highest (lowest) element of the data type. The *range* and the *reverse_range* attributes determine the index range. The *length* attribute means the number of elements of a bit_vector. For example,

```
subtype IDX_RANGE is INTEGER range 10 downto 0 ;
```

Types, Operators and Expressions 51

```
variable VEC1:   BIT_VECTOR (IDX_RANGE) ;
VEC1'left            -- equal to IDX_RANGE'left   == 10
VEC1'right           -- equal to IDX_RANGE'right  == 0
VEC1'high            -- equal to IDX_RANGE'high   == 10
VEC1'low             -- equal to IDX_RANGE'low    == 0
VEC1'range           -- equal to IDX_RANGE        == 10 downto 0
VEC1'reverse_range   -- equal to                  == 0 to 10
VEC1'length          -- 11
```

The *event* and *stable* attributes are associated with signals. It represents whether there is a change value to the signal at the current simulation time (or no change). In synthesis, they are used in association with the **wait** and **if** statements.

There are other predefined attributes. Most are used for simulation purposes. For example, the positional attributes such as *val, pos, succ, pred, leftof,* and *rightof,* the signal attributes such as *delayed, quite, active,* and *last_value,* and the simulation attributes such as *behavior, structure, last_event* and *last_active.*

3.4.6 Aggregates

An aggregate combines one or more values into a composite value of an array type or record type. It can be used to assign values to an object of array type or record type during the initial declaration or in an assignment statement. An element's index can be specified by using either *positional* or *named* association. By using positional association, each element is given the value of its expression in order. By using name association, the correspondence is made explicit by naming each element with a value.

For example,

```
    type Color_List  is (Red, Orange, Blue, White) ;
    type Color_Array is array (Color_List) of BIT_VECTOR(1 downto 0)
;
    variable X : Color_array ;
    X := ("00", "01", "10", "11") ; -- positional association
    X := (Red => "00",  Blue => "10", Orange => "01", White => "11")
;
```

It is not necessary to specify all element indices in an aggregate. The following shows two ways of using aggregates to assign values to a vector.

```
subtype BV7 is BIT_VECTOR(7 downto 0) ;
variable X : BV7 ;
-- the following three assignments produce the same result
X := ('0', '0', '0', '1', '1', '1', '1', '1') ;
X := ('0', '0', '0', others => '1') ;
X := BV7'('0', '0', '0', others => '1') ;  --qualified expression
```

3.4.7 Qualified Expressions

A qualified expression is an expression or an aggregate that is used to resolve ambiguities. It explicitly states the type or the subtype of an operand. A qualified expression is described as

type_name'(expression)

where *type_name* is the name of any defined type, and *expression* must evaluate to a value of an appropriate type.

For example, YELLOW is defined as an element in data type COLOR1 and data type COLOR2. On some occasions, just referring to YELLOW is ambiguous. We can add a qualifier to YELLOW to explicitly identify it data type.

```
type COLOR1 is (RED, ORANGE, YELLOW, GREEN, BLUE, PURPLE) ;
type COLOR2 is (GREEN, BLACK, WHITE, YELLOW) ;
function FOO return COLOR1 ;
function FOO return COLOR2 ;
if (FOO = YELLOW) then
endif;
-- Ambiguous FOO can be of type COLOR1 or COLOR2
if FOO = COLOR1'(YELLOW) then
endif;
-- qualified expression;
```

3.4.8 Type Conversions

A type conversion provides for explicit conversion between closely related types. Type conversions are different from qualified expressions because they change the type of the expressions, whereas qualified expressions simply resolve the type of an expression.

The syntax of the type conversion is

type_name(expression)

where *type_name* is the name of any defined type. The *expression* must evaluate to a value of a type that is *convertible* into type *type_name*. Explicit type conversions are allowed between closely related types. The example below shows a conversion of a signal of STD_LOGIC_VECTOR type to one in STD_ULOGIC_VECTOR type.

```
signal X: STD_LOGIC_VECTOR(3 downto 0) ;
signal Y: STD_ULOGIC_VECTOR(3 downto 0) ;
Y <= STD_ULOGIC_VECTOR(X) ;
```

3.5 SUMMARY

1. A named object may be a constant (unchanging value), a variable (changing value), or a signal (simulator time to schedule action).

2. VHDL is a strongly typed language. The integer 1, boolean true and the bit '1' are not the same.

3. In VHDL, an identifier must begin with an alphabetic character, followed by a letter, underscore, or digit. Identifiers are not case sensitive.

4. Literals are made up of characters or digits. Literals are case sensitive.

5. Expressions are formulas that define how to compute or qualify a value. Expressions perform arithmetic or logic computations by applying an operator to one or more operands.

Exercises

1. What does this mean in VHDL?

   ```
   signal X, A, B: BOOLEAN ;

   X  <= A  <= B;
   ```

2. In the following declaration, **A** is declared as a BIT type and B is declared as a BOOLEAN type.

   ```
   VARIABLE A: BIT;
   VARIABLE B: BOOLEAN;
   ```

   Write an expression to convert A to B. Assume '1' is true and '0' is false.

3. Which of the following statements are invalid?

   ```
   VARIABLE A, B, C, D: BIT_VECTOR ( 3 DOWNTO 0);
   VARIABLE E, F, G: BIT_VECTOR (1 DOWNTO 0);
   VARIABLE H, I, J, K: BIT;

   1) A := B XOR C AND D;
   2) H := I AND J OR K;
   3) H := I OR F;
   4) H := A(3) OR I;
   ```

4. Which of the following five statements are valid?

   ```
   SIGNAL A,B,C,D,E: IN BIT;
   SIGNAL OU: OUT BIT_VECTOR(3 downto 0);
   VARIABLE T: IN INTEGER;

   1)   T := A AND B;
   2)   OU <= A & B & C & D;
   3)   E <= NOT T;
   4)   OU <= E & OU(2 down to 0);
   5)   T := integer(B OR C);
   ```

Types, Operators and Expressions

5. Use the concatenation operation to perform a right shift operation on the following variable:

    VARIABLE A: BIT_VECTOR (31 downto 1);

6. Use the concatenation operation to perform an arithmetic right shift operation on the following variable:

    VARIABLE B: BIT_VECTOR (31 downto 1);

7. What are the values of the following constants:

    (a) 2#010_100#
    (b) 4#0_1_32#
    (c) 16#34_2#
    (d) 87

8. Given the following declaration:

    variable TEST: BIT_VECTOR(27 downto 0);

    What are the values of the following expressions?

    (a) TEST'left
    (b) TEST'right
    (c) TEST'high
    (d) TEST'low
    (e) TEST'range
    (f) TEST'reverse_range
    (g) TEST'length

9. Given the following declarations:

    type FOO is (monday, Wednesday, Friday) ;
    variable FOO1: BIT_VECTOR(10 downto 0) ;

    What is the meaning of the following expressions?

    (a) FOO'(monday)
    (b) FOO1'left

4
SEQUENTIAL STATEMENTS

An architecture determines the function of an entity. It consists of a declaration section where signals, types, constants, components, and subprograms are declared, followed by a collection of *concurrent statements*. The concurrent statements can be concurrent signal assignment statements, block statements and component instantiation statements. A concurrent statement communicates with other concurrent statements through signals. Each concurrent statement in an architecture defines a unit of computation that reads signals, perform computation based on their value, and assigns computed values to signals.

The process statement is one of the concurrent statements which is used to model the behavior of a circuit. All the processes within an architecture are executed concurrently. The statements within a process are *sequential statements*. Sequential statements in a process are executed as an infinitive loop starting from the first statement, second, third, until the last statement and go back to the first statement. The execution of a process is suspended by wait statements and reactivated by the change of the state of the signals in the sensitivity list. The following are the sequential statements defined in VHDL.

- Variable assignment statements
- Signal assignment statements
- If statements
- Case statements
- Null statements
- Assertion statements

- Loop statements
- Next statements
- Exit statements
- Wait statements
- Procedure calls
- Return statements

4.1 VARIABLE ASSIGNMENT STATEMENTS

A variable assignment statement assigns a new value specified by an expression to a target variable. The syntax is

  target_variable := *expression* ;

The left side of the variable assignment statement is a variable object previously declared. The right side of the variable assignment is an expression using variables, signals, constants, and literals. The base type of the *expression* must be the same as that of the *target_variable*.

When a variable is assigned, the assignment executes in zero simulation time. In other words, it changes the value of the variable immediately at the current simulation time. Variables can only be declared in a process or a subprogram. They are used as temporary storages and cannot be seen by other concurrent statements. For example,

```
subtype INT16 is INTEGER range 0 to 65535 ;
signal S1, S2           : INT16 ;
signal GT               : BOOLEAN ;
process(S1, S2)
   variable  A, B       : INT16 ;
   constant  C          : INT16 := 100 ;
begin
    A  := S1 + 1 ;
```

Sequential Statements

```
       B  := S2 * 2 - C ;
       GT <= A > B ;
    end process ;
```

4.2 SIGNAL ASSIGNMENT STATEMENTS

A signal assignment changes the value of a signal. Signals are associated with time. When a signal is assigned a value, the assignment will not take effect immediately, instead will be scheduled to a future simulation time. The syntax is

target_signal <= [**transport**] *expression* [**after** *time_expression*] ;

where the *expression* determines the assigned value and the base type of the expression must be the same as that of the *target_signal*. The *time_expression* in the **after** clause must be of the predefined type TIME.

Within a process, the assignment to a signal will be delayed until a simulation cycle is run, triggered by a WAIT statement. Take the following program as an example.

```
       process
       begin
           S1 <= not CLK after 30 ns ;
           S2 <= Data_In after 0 ns ;
           S3 <= S1 and S2 ;
           wait on CLK ;
       end ;
```

A transaction of assigning the complement of CLK to S1 will be scheduled at the 30 ns after the current simulation time. A transaction of assigning the current value of Data_In to S2 will be scheduled at a delta time after the current simulation time. Similarly, a transaction fo assigning the "and" of the current value of S1 and S2 will be assigned to signal S3. When an event occurs (CLK changes from '0' to '1'), all the transactions scheduled before the time of the event will be evaluated by the simulator.

When a signal is assigned a value in a process, it defines a *driver* of the signal. Within a process, each signal should have only a single driver. If a signal is assigned value in multiple processes, it is said to have multiple drivers. In VHDL, a resolution function should be defined to resolve a signal with multiple drivers.

There are two types of delay that can be applied when scheduling a time/value pair into the driver of a signal: *inertial* and *transport*. Inertial delay is the default in VHDL. It is used for devices that do not respond unless a value on its input persists for the given amount of time. Inertial delay is useful in modeling devices that ignore spikes on their inputs. Transport delay is analogous to the delay incurred by passing a current through a wire.

The signal assignments using inertial delay and transport delay have different effects during the simulation. Consider the following process:

```
signal S: BIT := '0';
process
begin
  S <= '1' after 5 ns ;
  S <= '0' after 10 ns ;
  wait on CLK ;
end process ;
```

The first statement schedules a transaction of assigning '1' to S at the 5 ns after the current simulation time. The second assignment overrides the first assignment because a new value is introduced to the driver with inertial delay. In other words, the first assignment is redundant. One might think of the second assignment as saying S must have an inertia of 10 ns.

If we reverse the order of two assignments of the previous process as shown in the following program segment, the later one will still override the first one. In other words, only the transaction of assigning '1' to S at the 5 ns after the current simulation time will be scheduled.

```
signal S: BIT := '0';
process
begin
  S <= '0' after 10 ns ;
  S <= '1' after 5 ns ;
  wait on CLK ;
end process ;
```

Sequential Statements

If the signal delay uses transport delay, the interpretation will be different.

```
signal S: BIT := '0';
process
begin
    S <= transport '1' after 5 ns ;
    S <= transport '0' after 10 ns ;
    wait on CLK ;
end process ;
```

The first statement schedules a transaction of assigning '1' to S at the 5 ns after the current simulation time. The second statement adds another transaction to the driver for S. In other words, there will be two transactions scheduled to be evaluated. One is after 5 ns, and the other is after 10 ns of the current simulation time.

If we reverse the order of the two statements, the process has different effect. The first assignment will place a transaction of assigning '0' to S at the 10 ns after the current simulation time. When the second assignment is executed, this transaction will be overwritten by the second transaction which is assigning '1' to S at the 5 ns after the current simulation time.

In the following, we show some more examples to illustrate the difference between the signal assignments and variable assignments. For example, in the following program segment, signal S1 is assigned a value of '1' and then assigned a value of '0' before the wait statement. The second assignment will override the first assignment. Therefore S1 will get the value of '0' after the wait statement is evaluated.

```
signal S1,CLK   : BIT;
Main : process
begin
    S1 <= '1';
    S1 <= '0';
    wait until CLK'event and CLK = '1' ;   -- leading edge
end process Main;
```

In the next example, a transaction of assigning 10 to S1 will be scheduled at a delta time after the current simulation time, and a transaction of assigning the current value of S1(which is 5) to S2 will be scheduled at a delta time after the current simulation time. Therefore, after the second **wait** statement, S1 and S2 will be assigned to 10 and 5, respectively.

```
signal S1, S2    :    INTEGER ;
Main : process
begin
   S1 <= 5 ;
   wait until CLK'event and CLK = '1' ;
   S1 <= 10 ;
   S2 <= S1 ;        -- S2 gets a value of 5, not 10
   wait until CLK'event and CLK = '1' ;
end process Main ;
```

In the next example, when the statement, "V1 := 10 ;", is evaluated, the value of the signal V1 is assigned immediately. Then, a transaction of assigning the new value of V1 to S2 will be scheduled. Hence, after the second **wait** statement, S2 will be assigned a value of 10.

```
signal S2         :    INTEGER ;
Main : process
   variable V1 :    INTEGER;
begin
   V1 := 10 ;
   S2 <= V1 ;        -- S2 gets a value of 10
   wait until CLK'event and CLK = '1' ;
end process Main ;
```

4.3 IF STATEMENTS

An **if** statement creates a branch in the execution flow. According to the evaluation of the conditions, one or none of the enclosed sequences of statements will be executed. The syntax is

>**if** *condition* **then**
>   { *sequential_statement* }
>{ **elsif** *condition* **then**
>   { *sequential_statement* } }

Sequential Statements

```
        [ else
            { sequential_statement } ]
        end if;
```

An expression specifying a *condition* must be a BOOLEAN type expression. Each branch of an **if** statement can have one or more *sequential_statement*. The *condition* of the **if** statement is first evaluated. If the *condition* is TRUE, then the statement immediately following the keyword "then" is executed; else the *conditions* following **elsif** clauses (treating a final **else** as **elsif** TRUE **then**) are evaluated in succession. If one *condition* is TRUE, then the corresponding sequence of statements is executed; otherwise, execution continues immediately with the statement after the **if** statement.

The following example can be used to represent a 2-input logical **and** gate.

```
signal IN1, IN2, OU   : STD_LOGIC ;
And_Process :  process (IN1, IN2)
begin
   if IN1 = '0' or IN2 = '0' then
      OU <= '0';
   elsif IN1 = 'X' or IN2 = 'X' then
      OU <= 'X';
   else
      OU <= '1';
   end if;
end process ;
```

4.4 CASE STATEMENTS

The case statement is another form of conditional control provided in VHDL. It is a multiway branch based on the value of a control expression. One of a number of alternative sequences of statements will be executed when the control expression is equal to that of the chosen alternative. It is like the if statement in that a condition is used to choose among a number of collections of statements. It is different however, in that the statements are chosen based on the value of an expression.

The syntax is

```
case expression is
    when choices =>
        { sequential_statement }
    { when choices =>
        { sequential_statement } }
end case;
```

The *expression* must evaluate to be an integer type, an enumerated type, or a one-dimensional character array type such as BIT_VECTOR. Each *choice* can be either a static expression or a static range. Each value in the range of *expression*'s type must be covered by one and only one choice.

The final choice can be **others**, which match all remaining choices in the range of expression's type. The following restrictions are placed on the choices:

1. Each choice must be of the same type as the *expression*.

2. Each value must be represented once and only once in the set of choices of the case statement.

3. If no **others** choice is presented, all possible values of the expression must be covered by the set of choices.

The following shows an example of a case-statement.

```
signal S1          : INTEGER range 0 to 7 ;
signal I1, I2, I3 : BIT ;
select_process :
process(S1, I1, I2, I3)
begin
    case S1 is
        when 0 | 2 =>
            OU <= '0' ;
        when 1 =>
            OU <= I1 ;
        when 3 to 5 =>
            OU <= I2 ;
        when others =>
```

Sequential Statements 65

```
              OU <= I3 ;
    end case ;
end process select_process ;
```

4.5 NULL STATEMENTS

The syntax of null statement is

   null ;

There is no action for a *null* statement in VHDL. The system will ignore the null statement and proceed to the next statement. This statement is usually used to explicitly state that no action is to be performed when a condition is true. Since the choices in a case statement have to cover all possible values of the case expression, the null statement can be used for those choices which need no action. For example,

```
variable SEL : INTEGER range 0 to 31;
variable V   : INTEGER range 0 to 31;
case SEL is
   when 0 to 15  =>
      V := SEL ;
   when others  =>
      null ;         -- specify no action here
end case;
```

4.6 ASSERTION STATEMENTS

The assertion statement is used to indicate a certain condition is expected to be true. If the condition is false, it reports a diagnostic message. The syntax is

            assert *condition*
              [**report** *error_message*]
              [**severity** *severity_expression*] ;

where the *condition* must be a BOOLEAN type expression. The *error_message* is a STRING type expression and usually contains information and position of the assertion; the *severity_expression* is an expression of predefined type SEVERITY_LEVEL which specifies the severity level of the assertion.

The assertion statement is useful for detecting condition violation in VHDL simulation cycle. There is no action as far as the synthesis tool is concerned similar to the **null** statement. For example,

```
subtype COUNTER_TYPE is INTEGER range 0 to 65535;
variable Time_Counter,Suspension_Counter :  COUNTER_TYPE;
constant Time_Out :  COUNTER_TYPE := 10000;
assert ((Time_Counter - Suspension_Counter) > Time_Out)
    report "Timing Violation in delay function"
    severity ERROR ;
```

4.7 LOOP STATEMENTS

A **loop** statement contains a loop body that is to be executed repeatedly, zero or more times. The loop body is a sequence of statements encapsulated by the loop statement. The syntax is

> [*label* :] [*iteration_scheme*] **loop**
>   { *sequential_statements* } |
>   { **next** [*label*] [**when** *condition*] ; } |
>   { **exit** [*label*] [**when** *condition*] ; }
> **end loop** [*label*] ;

where the optional *label* names the loop and is useful for building nested loops. Each loop statement has a range bounded by the **end loop**. If a *label* name appears after the **end loop**, it must describe the same label at the beginning of a loop. The following discuss the iteration schemes.

A **loop** statement without an *iteration_scheme* describes the repeated execution of the sequence of statements. The execution of the loop statement is complete when the control is transferred out of the loop body by an **exit** statement. Also, the **next** statement can modify the execution of the loop statement. For example,

Sequential Statements

```
    Count_Down:  process
       variable Min, Sec :   INTEGER range 0 to 60 ;
    begin
       L1 :  loop
          L2 :  loop
             exit L2 when (Sec = 0) ;
             wait until CLK'event and CLK = '1' ;
             Sec := Sec - 1 ;
          end loop L2 ;
          exit L1 when (M = 0) ;
          Min := Min - 1 ;
          Sec := 60 ;
       end loop L1 ;
    end process Count_Down;
```

Another iteration scheme of the **loop** statement is the **for** loop. A **for** loop is a sequential statement in a process that iterates over a number of values.

```
    for I in 1 to 10 loop
       I_SQUARED(I) := I * I ;
    end loop ;
```

In this example, the **for loop** iterates 10 times. Its function is to calculate the squares from 1 to 10 and insert them into array I_SQUARED. The loop index I does not need to be declared, nor can it be reassigned a value within the loop. It is by default integer and takes on the values 1 to 10.

A *downto* clause may be used to create a descending range. For example,

```
    for I in X downto Y loop
       I_SQUARED(I) := I * I ;
    end loop ;
```

The last iteration scheme is the **while** loop, where the statement is executed by first evaluating the *condition*. If the result is TRUE, then the loop body is executed. The entire process is then repeated, alternately evaluating the *condition* and then, if the value is TRUE, executing the loop body. The execution of the loop statement will be terminated when the *condition* evaluates to FALSE, or when control is transferred out of the loop body by an **exit** statement. Also, the **next** statement can modify the execution of the while loop statement. For example,

```
process
    variable A, B, C, D : INTEGER ;
begin
    . . .
    while ((A + B) > (C + D)) loop
        A := A - 1 ;
        C := C + B ;
        next when (B < 10) ;
        B := B - D ;
    end loop ;
    . . .
end process ;
```

For loop statements, there must be at least one wait statement in each (enclosed) logic branch which contains a signal assignment. Otherwise, the synthesis and the simulation waveform may differ. In the next example, signal S is updated 10 times within the loop. This behavior will not show up in simulation. However, the intermediate results may show up during the simulation of a synthesis result.

```
signal S       : INTEGER range 0 to 10 ;
process
    variable I : INTEGER range 0 to 10 ;
begin
    wait until (CLK'event and CLK = '0');
    I := 0 ;
    while (I < 10) loop
        S <= I ;
        I := I + 1 ;
    end loop ;
end process ;
```

4.8 NEXT STATEMENTS

The **next** statements are sequential statements used only within loops. The **next** statement skips execution to the next iteration of an enclosing loop statement (called *loop_label* in the syntax). The completion is conditional if the statement includes a condition.

The syntax is:

   next [*loop_label*] [**when** *condition*] ;

Execution of the **next** statement causes execution of the body of the (labeled, if any) loop statement to be terminated and transfers the control to the next loop index value. When loops are nested, each loop may be given a label. The next statement refers to a particular loop label to skip. If the *loop_label* is absent, the next statement applies to the innermost enclosing loop.

The following example shows the use of a **next** statement.

```
L1 :  while I < 10 loop
   L2 :  while J < 20 loop
         .
         .
         next L1 when I = J ;
         .
      end loop L2 ;
end loop L1 ;
```

4.9 EXIT STATEMENTS

Similar to the **next** statement, the **exit** statement can also be used within loops. The **exit** statement skips the remainder of an enclosing loop statement (called *loop_label* in the syntax) and continues with the next statement after the exited loop. The completion is conditional if the statement includes a condition.

The syntax is:

exit [*loop_label*] [**when** *condition*] ;

The *loop_label* in the exit statement identifies the particular loop to be exited. If the *loop_label* is absent, the exit statement applies to the innermost enclosing loop. The next example shows the use of an **exit** statement.

```
Natural_Modulus :   loop
   B := B - A ;
   exit when B > A ;
end Natural_Modulus ;
```

4.10 WAIT STATEMENTS

The **wait** statement causes a simulator to suspend execution of a process statement or a subprogram, until some conditions are met. The objects being waited upon should be signals. The conditions for resuming execution of the suspended process or subprogram can be specified by three different means. The syntax is:

> **wait**
>   [**on** *signal_name* {*signal_name* }]
>   [**until** *boolean_expression*]
>   [**for** *time_expression*] ;

The **wait on** signal clause specifies a list of one or more signals upon which the **wait** statement will wati for events. In the following example, the simulator suspends execution of the process until an event (change) occurs on either signal A or B. Then, execution will continue with the statement following the **wait** statement.

```
wait on A, B ;
```

The **wait until** *boolean_expression* clause will suspend execution of the process until the expression returns a value of true. This statement will create an implicit sensitivity list of the signals used in the expression. When any of the signals in the expression have events occur upon them, the expression will be evaluated. If the until clause is absent, "wait until true" is assumed. In the

Sequential Statements

following example, the simulator suspends execution of the process until an event occurs to signal x and condition $x < 10$ becomes satisfied.

```
wait until x < 10 ;
```

The **wait for** *time_expression* clause will suspend execution of the process for the time specified by the time expression. After the time specified in the time expression has elapsed, execution will continue on the statement following the **wait** statement. If the for clause is absent, then "wait for time'high" is assumed. In other words, no explicit time is specified to bind the wait statement. For example,

```
wait for 10 ns ;
```

The different options of a **wait** statement can be used together. For example,

```
wait on A,B until (X < 10) for 10 ns ;
```

This statement will wait for an event on signals A and B and will continue only if $(X < 10)$ becomes true at the time of the event, or until 10 ns of time has elapsed.

For synthesis, the wait statements are commonly used to specify clock inputs. Synthesis tools usually support the wait statement in a limited form. For details, please refer to Section 10.2 of Chapter 10. In a VHDL design description, a **wait** statement suspends a process until a positive-edge or negative-edge is detected on a signal.

For example, the following program shows the behavior of a D-type flip-flop. The value of signal D is assigned to signal Q at every leading edge of signal CLK.

```
entity D_FF is
   port( CLK : in      BIT ;
         D   : in      BIT ;
         Q   : buffer BIT ) ;
end D_FF ;
architecture BHV of D_FF is
begin
   process
   begin
      wait until CLK'event AND CLK = '1' ;
```

```
        Q <= D ;
     end process ;
  end BHV ;
```

The following shows how a **wait** statement is used to describe a circuit which suspends the process until the falling edge of clock and then signal START becomes TRUE.

```
   signal START    : BOOLEAN ;
   process
      .  .  .
   begin
      wait until CLK'event and CLK = '0' and START ;
      .  .  .
   end process ;
```

4.11 PROCEDURE CALLS

In a behavior design description, subprograms provide a convenient way of documenting frequently used functions. There are two different types of subprograms: a *procedure*(returns multiple values) and a *function*(returns a single value). A subprogram is composed of sequential statements just like a process.

Procedure calls invoke procedures to be executed in a process. The procedure must be declared in a declaration section prior to its call. These procedures can be thought of as extensions to the behavior of the process statement. They serve two purposes. First, a procedure can be written to isolate complicated sections of a process statement. This provides a further functional breakdown of the behavior and makes the process statement easier to read. Procedures are also used to encapsulate behavior code which can be used by different processes. The behavior of the procedure is the same for all calls but the parameters may be different.

4.12 RETURN STATEMENTS

The **return** statement terminates a subprogram. The return statement can only be described within a function or a procedure. It is required in a function body, but optional in a procedure body. The syntax is

> **return** [*expression*] ;

where *expression* provides the function's return value. The return statement within a function must have an *expression* as it's return value, but the return statement appeared in procedures must not have the *expression*. The expression's type must match the return type of declared function. Every function body must have at least one return statement. A function can have more than one return statement. Only one return statement is reached by a given function call.

4.13 SUMMARY

1. A *process* defines regions in architectures where sequential statements are executed.

2. Process statements provide concurrent processing capability using local variables and global signals.

3. Variable assignment statements are executed in zero simulation time. The execution of a signal assignment is delayed until a simulation cycle is run, triggered by the execution of a WAIT statement.

4. WAIT statements dynamically control process suspension/execution. In simulation, all processes are started and executed up to WAIT statements.

5. A process can call *functions*(that return a single value) and *procedures* (that return more than one value).

Exercises

1. Write down the value of s1, s2 and s3 for the first three clock cycles

   ```
   signal S1, S2, CLK: BIT:= '0';
   begin
      process
      variable V1:  BIT := '0';
      begin
         V1 := not V1;
         S1 <= V1;
         S2 <= S1;
         wait until CLK'event and CLK='1';
      end process;
   end;
   ```

2. There are three errors in the following program. Circle them out.

   ```
   SIGNAL s1, clk: BIT;
   begin
      process(clk)
      VARIABLE v1: BIT;
      begin
         v1 <= not s1;
         s1 := v1;
         wait until clk'event and clk='1';
      end process;
   end;
   ```

3. Use a CASE statement to describe the behavior of a 2-to-1 multiplexor.

4. Use a CASE statement to describe the behavior of a 2-to-4 decoder.

5. Use an IF statement to describe the behavior of a 2-to-4 decoder.

6. Assume you have an 8-bit ALU which executes 4 functions: addition when the control input is 00, subtraction when the control input is 01, AND when the control input is 10, and OR when the control input is 11. Write a VHDL description whose behavior is the same as the above mentioned ALU.

5
CONCURRENT STATEMENTS

An architecture which describes the functionality of an entity consists of one or more concurrent statements. Each concurrent statement defines a unit of computation that reads signals, perform computation based on their value, and assigns computed values to signals. They all together define interconnected components and processes that jointly describe the overall structure and/or behavior of a design entity. The concurrent statements are executed in parallel. It does not matter on the order they appear in the architecture.

The concurrent statements can be concurrent signal assignment statements, block statements, component instantiation statements, and generate statements. We will discuss the following concurrent statements in this chapter. The discussion of component instantiation statements and the generate statement will be discussed in chapter 7.

- Process statements
- Concurrent signal assignments
- Conditional signal assignments
- Selected signal assignments
- Block statements
- Concurrent procedure calls
- Concurrent assertion statements

5.1 PROCESS STATEMENTS

A process is composed of a set of sequential statements, but processes are themselves concurrent statements. All the processes in a design execute concurrently. However, at any given time only one sequential statement is executed within each process. A process communicates with the rest of a design by reading or writing values to and from signals or ports declared outside the process.

The syntax for a process statement is

> [*label* :] **process** [(*sensitivity_list*)]
>   { *process_declaration_part* }
> **begin**
>   { *sequential_statements* }
> **end process** [*label*] ;

A *process_declaration_part* defines objects that are local to the process, and can have any of the following items:

- variable declaration
- constant declaration
- type declaration
- subtype declaration
- subprogram body
- alias declaration
- use clause

The statements allowed in the *sequential_statements* are described in chapter 4.

If a sensitivity list appears following the reserved word **process**, then it has the same meaning as a process containing a wait statement as the last statement and is interpreted as

> **wait on** *sensitivity_list*;

Concurrent Statements

where the *sensitivity_list* of the wait statement is the same as that following the reserved word **process**. Such a process statement need not contain any explicit **wait** statement.

The execution of a process statement consists of the repeated execution of its sequential statements. After the last statement in a process is executed, the execution once again begins from the first of the sequential statements in a process. The behavior of the process is like a pseudo infinite loop statement which encloses the whole sequential statements specified in the process. The execution of a process is suspended by a wait statement and may be reactivated when there is an event occurs to the signals in the sensitivity list.

The following example shows a process description of an 8-to-3 priority encoder. The process has a sensitivity list consisting of signals Y1, Y2, Y3, Y4, Y5, Y6, and Y7 which is specified following the process declaration. It is equivalent to a process having a **wait on Y1, Y2, Y3, Y4, Y5, Y6, Y7;** at the last statement of the process. Whenever there is an event occurs to any of the signals in the sensitivity list, the process will be re-evaluated.

```
entity PRIORITY is
    port ( Y1, Y2, Y3, Y4, Y5, Y6, Y7 :  in BIT ;
           VEC : out BIT_VECTOR(2 downto 0)    ) ;
end PRIORITY ;
architecture BEHAVIOR of PRIORITY is
begin
    process (Y1, Y2, Y3, Y4, Y5, Y6, Y7)
    begin
        if    (Y7 = '1') then VEC <= "111" ;
        elsif(Y6 = '1') then VEC <= "110" ;
        elsif(Y5 = '1') then VEC <= "101" ;
        elsif(Y4 = '1') then VEC <= "100" ;
        elsif(Y3 = '1') then VEC <= "011" ;
        elsif(Y2 = '1') then VEC <= "010" ;
        elsif(Y1 = '1') then VEC <= "001" ;
        else                  VEC <= "000" ;
        end if ;
    end process ;
end BEHAVIOR ;
```

The next example shows a description of two concurrent communicating processes. The Send process has a sensitivity list consisting of signals CLK and ACK, and the Receive process has a sensitivity list consisting of signals CLK and READY. The two processes synchronize their operations using ACK and READY signals.

```
entity HANDSHAKE is
   port ( CLK      : in   BIT ;
          DATA_IN  : in   INTEGER ;
          DATA_OUT : out  INTEGER ) ;
end HANDSHAKE ;
architecture PROTOCOL of HANDSHAKE is
   signal READY, ACK : BIT ;
   signal DATA       : INTEGER ;
begin
   Send : process
   begin
      READY <= '1' ;
      DATA  <= DATA_IN ;
      wait until CLK'event and CLK = '1' and ACK = '1' ;
      READY <= '0' ;
      wait until CLK'event and CLK = '1' and ACK = '0' ;
   end process ;
   Receive : process
   begin
      ACK <= '0' ;
      wait until CLK'event and CLK = '1' and READY = '1' ;
      DATA_OUT <= DATA ;
      ACK <= '1' ;
      wait until CLK'event and CLK = '1' and READY = '0' ;
   end process ;
end PROTOCOL ;
```

Concurrent Statements 79

5.2 CONCURRENT SIGNAL ASSIGNMENTS

Another form of a signal assignment is a concurrent signal assignment, which is used outside a process, but within an architecture. The simplest form of the concurrent signal assignment is

>   *target* <= *expression* [**after** *time_expression*] ;

where *target* is a signal that receives the value of *expression*. Similar to the sequential signal assignment, the **after** clause is ignored by the synthesizer.

A concurrent signal assignment statement represents an equivalent process statement that assigns values to signals. The following example shows two concurrent signal assignments within an architecture.

```
architecture Description1 of EXAMPLE is
signal I1, I2, I3, I4, AND_OUT, OR_OUT: BIT ;
   . . .
begin
    AND_OUT <= I1 and I2 and I3 and I4 ;
    OR_OUT  <= I1 or I2 or I3 or I4 ;
end Description1 ;
```

The above example is equivalent to the VHDL program in the following.

```
architecture Description2 of EXAMPLE is
signal I1, I2, I3, I4, AND_OUT, OR_OUT: BIT ;
   . . .
begin
   process (I1, I2, I3, I4)
   begin
       AND_OUT <= I1 and I2 and I3 and I4 ;
   end process ;
   process (I1, I2, I3, I4)
   begin
       OR_OUT <= I1 or I2 or I3 or I4 ;
   end process ;
end Description2 ;
```

The above example is also equivalent to the VHDL program in the following.

```
architecture Description3 of EXAMPLE is
signal I1, I2, I3, I4, AND_OUT, OR_OUT: BIT ;
   . . .
begin
   process
   begin
      AND_OUT <= I1 and I2 and I3 and I4 ;
      wait on I1, I2, I3, I4 ;
   end process ;

   process
   begin
      OR_OUT <= I1 or I2 or I3 or I4 ;
      wait on I1, I2, I3, I4 ;
   end process ;
end Description3 ;
```

5.3 CONDITIONAL SIGNAL ASSIGNMENTS

A conditional signal assignment is a concurrent statement and has one target, but can have more than one expression. Except for the final expression, each expression goes with a certain condition. The conditions are evaluated sequentially. If one condition evaluates to TRUE, then the corresponding expression is used; otherwise the remaining expression is used. One and only one of the expressions will be used at a time.

The syntax for the conditional signal assignment is

$target <= \{ expression\ [\ \textbf{after}\ time\_expression\ ]\ \textbf{when}\ condition\ \textbf{else}\ \}$
$expression\ [\ \textbf{after}\ time\_expression\ ]\ ;$

Concurrent Statements

Any conditional signal assignment can be described by a process statement which contains an **if** statement. The following shows an example of a conditional signal assignment statement.

```
architecture Description1 of EXAMPLE is
  signal A, B, X, Z: BIT ;
  constant C: BIT :='0' ;
  . . .
  Z <= A when ( X > 10 ) else
       B when ( X > 5  ) else
       C ;
end Description1 ;
```

The statement is equivalent to the following process.

```
architecture Description2 of EXAMPLE is
  signal A, B, X, Z: BIT ;
  constant C: BIT :='0' ;
  . . .
  process (A, B, X)
  begin
    if ( X > 10 ) then
       Z <= A ;
    elsif ( X > 5 ) then
       Z <= B ;
    else
       Z <= C ;
    end if ;
  end process ;
end Description1 ;
```

The **after** clauses in conditional signal assignment are ignored by the synthesis tool.

5.4 SELECTED SIGNAL ASSIGNMENTS

A selective signal assignment can have only one target and can have only one **with** expression. This value is tested for a match in a manner similar to the **case** statement. It runs whenever any change occurs to the selected signal.

The syntax is

>   **with** *choice_expression* **select**
>   *target* <= { *expression* [**after** *time_expression*] **when** *choices* , }
>   *expression* [**after** *time_expression*] **when** *choices;*

The following shows an example of a selected signal assignment statement.

```
with SEL select
    Z <= A when 0 | 1 | 2 ,
         B when 3 to 10,
         c when others ;
```

Any selected signal assignment can be described by a process statement which contains a case statement. For example, the above statement is equivalent to the following process.

```
process (SEL, A, B, C)
begin
    case SEL is
        when 0 | 1 | 2 =>
            Z <= A ;
        when 3 to 10 =>
            Z <= B ;
        when others =>
            Z <= C ;
    end case ;
end process ;
```

5.5 BLOCK STATEMENTS

A block statement consists of a set of concurrent statements. The statement area in an architecture can be broken down into a number of separate logical areas. Each block represents a self-contained area of the model. Blocks are used to organize a set of concurrent statements hierarchically.

The syntax is

> *label* : **block**
>   { *block_declarative_part* }
> **begin**
>   { *concurrent_statement* }
> **end block** [*label*] ;

A *block_declarative_part* defines objects that are local to the block, and can have any of the following items:

- signal declaration
- constant declaration
- type declaration
- subtype declaration
- subprogram body
- alias declaration
- component declaration
- use clause

The order of each *concurrent_statement* within a block is not significant, since each statement is always active. *Concurrent_statements* pass information through signals.

Objects declared in a block are visible to that block, and to all blocks nested within. When a child block declares an object with the same name as the one in the parent block, the child's declaration overrides that of the parent.

In the next example, block B1_1 is nested within block B1. Both B1 and B1_1 declare a signal named S. According to the scoping rule, the signal S used in block B1_1 will be the one declared within block B1_1, while the S used in block B2 is the one declared in B1.

```
architecture BHV of EXAMPLE is
    signal OU1     : INTEGER ;
    signal OU2     : BIT     ;
begin
    B1 : block
        signal S   : BIT ;
    begin
        B1_1 : block
            signal S : INTEGER ;
        begin
            OU1 <= S;
        end block B1_1 ;
    end block B1 ;
    B2 : block
    begin
        OU2 <= S;
    end block B2;
end BHV ;
```

5.6 CONCURRENT PROCEDURE CALLS

A concurrent procedure call represents a process containing the corresponding sequential procedure call. For each concurrent procedure call, there is an equivalent process statement. There is an equivalent process statement has no sensitivity list, an empty declarative part, and a statement part that consists of a procedure call statement followed by a wait statement. The function call statement consists of the same procedure name and actual parameter part that appear in the concurrent function call. Each formal parameter of a procedure that is invoked by a concurrent function call must be of class constant or signal.

Concurrent Statements

Since the block statement is used to specify the hierarchy of a circuit design, not for describing the behavior of the circuit in MEBS, the guarded block (a block with *guard_expression* and GUARD signal) in standard VHDL is not supported by MEBS to model the circuit, so the guarded assignments (the concurrent signal assignments with the option **guarded**), the guarded signals (the signals declared with a signal kind, **register** or **bus**), the disconnect specifications and null waveform element (both are used to turn off the driver of guarded signals) are also not supported. Furthermore, MEBS does not currently support *block_header* (generic and port interface for a block).

5.7 CONCURRENT ASSERTION STATEMENTS

The concurrent assertion statement performs the same action and is used for the same reason as the sequential assertion statements within a process. This statement is used for simulation purpose and will be ignored by the synthesis tool.

5.8 SUMMARY

1. Concurrent statements are executed in parallel. The order of them in a process is unimportant.

2. All the processes are executed concurrently. However, at any given time, only one sequential statement is interpreted with each process.

3. A sensitivity list is equivalent to having a WAIT at the end of the process, not at the beginning of the process. WAIT statements can be used to dynamically control process suspension and execution.

4. Concurrent signal assignments, conditional signal assignments, and selective signal assignments are shorthand notation of processes.

5. A block statement consists of a set of concurrent statements. Blocks are used to organize a set of concurrent statements hierarchically.

Exercises

1. Write a 7 segment display decoder using selected signal assignment.

2. Describe a 4-to-1 multiplexor using

   (a) a concurrent selected signal assignment

   (b) a concurrent conditional signal assignment

   (c) a concurrent signal assignment

3. Describe a 2-to-4 decoder using

   (a) a concurrent selected signal assignment

   (b) a concurrent conditional signal assignment

   (c) a concurrent signal assignment

4. Use concurrent signal assignments to check if two 4-bit numbers are equal. Assume the inputs A and B are type BIT_VECTOR and the output EQ1 is a BOOLEAN type object.

5. Assume the clock cycle time for the following circuit is 50 ns. Draw the timing diagram for the following description.

```
architecture SEQUENTIAL of EX is
   signal A, B, C: BIT := '0';
begin
  process
  begin
    A <= '1' ;
    B <= not (A or C) ;
    C <= not B ;
    wait until clock'event and clock = '1';
  end process ;
end SEQUENTIAL ;
```

6. Draw the timing diagram for the following description.

```
entity RSFF if
   port (S, R   : in BIT ;
```

```
               Q, QB   : buffer BIT );
end RSFF ;
architecture FOO of RSFF is
signal S  : BIT := '0' ;
signal R  : BIT := '1' ;
begin
  Q  <= S nand QB ;
  QB <= R nand Q ;
end FOO;
```

6

SUBPROGRAMS AND PACKAGES

In this chapter, subprograms and packages will be discussed. Subprograms consists of procedures and functions that can be invoked repeatly from different locations in a VHDL description. Packages define all the names of items that may be shared among several entities.

6.1 SUBPROGRAMS

The predefined operators, such as +, *, <, xor, *etc.* , require a fixed number of operands whose types must conform to certain rules. In case a user wants to define complex operators which will be used in many locations of a program, he/she can use subprograms. A subprogram can be seen as a generalized operator. It defines a function using a sequence of declarations and statements in behavior constructs and can be invoked from different locations in a VHDL program. Subprograms offer the possibility of describing complex descriptions succinctly.

VHDL provides two kinds of subprograms: *procedures* and *functions*. A procedure is invoked as a statement and a function is invoked as an expression. A procedure can return none, one, or more than one argument, while a function always returns just one argument. Procedures are permitted to change the values of the objects associated with the procedure formal parameters; therefore, parameters of procedures may be of mode **in**, **out**, and **inout**. Functions are intended to be used strictly for computing values and not for changing the value of any objects associated with the function's formal parameters; therefore, all parameters of functions must be of mode **in** and must be of class **signal** or

constant. A function must declare the type of the value the function returns. ormal parameters

The description of a subprogram can be divided into subprogram declaration and subprogram body. The subprogram declaration contains only interface information, while the subprogram body describes its functionality. The syntax for a subprogram declaration consists of a *subprogram-specification* followed by a semicolon, where the *subprogram-specification* is defined as:

subprogram-specification ::=
    procedure *identifier interface_list* |
    function *identifier interface_list* **return** *type-mark*

The *identifier* defines the name of the subprogram, and the *interface_list* defines the formal parameters of the subprogram. Each parameter is specified using the following syntax:

    [class] *name_list* [mode] *type_name* [:= *expression*] ;

where the class of a object refers to **constant, variable**, or **signal**, and the mode of a object can be **in, out**, or **inout**.

If no mode is specified for a subprogram parameter, the parameter is interpreted as having mode **in**. If no class is specified, parameters of mode **in** are interpreted as class **constant**, and parameters of mode **out** and **inout** are interpreted as being of class **variable**.

The following are examples of subprogram declarations:

```
            function Incrementer (Count:   INTEGER) return INTEGER ;
                        -- by default will be understood as:
                        -- constant Count:   in INTEGER
            procedure Send ( signal CLK    : in BIT ;
                             Data   :  INTEGER ;
                        -- by default will be understood as:
                        -- constant Data :   in INTEGER;
                             signal Ack    :  BIT ;
```

Subprograms and Packages

```
                    signal Ready  :  out BIT ;
                    signal Wire   :  out INTEGER ) ;
```

The syntax of a subprogram (procedure or function) body is

subprogram-specification **is**
    { *subprogram_declarative_item* }
begin
    { *sequential_statement* }
end [*identifier*] ;

The following shows examples of subprogram (procedure and function) bodies. Note that when a function is called, the function call must terminate by executing a **return** statement, which determines the value returned by the function call.

```
function Incrementer (Count :  INTEGER) return INTEGER is
   variable TEMP: INTEGER ;
   begin
      if (Count >= 255) then
         TEMP := 0 ;
      else
         TEMP := Count + 1 ;
      end if ;
      return(TEMP) ;
end Incrementer ;
procedure Send ( signal CLK    :  BIT ;
                        Data   :  INTEGER ;
                 signal Ack    :  BIT ;
                 signal Ready  :  out BIT ;
                 signal Wire   :  out INTEGER ) is
begin
   wait until (CLK'event and CLK = '1' and Ack = '0') ;
   Wire  <= Data ;
   Ready <= '1' ;
   wait until (CLK'event and CLK = '1' and Ack = '1');
```

```
        Ready <= '0';
end Send ;
```

In a subprogram call, the actual designator associated with a formal parameter of class **variable** must be a variable; The actual designator associated with a formal parameter of class **signal** must be a signal; and the actual designator associated with a formal parameter of class **constant** must be an expression. For **constant** or **variable** parameters, the values are transferred into or out of a subprogram by call-by-value. For **signal** parameters, the value is passed by reference. In other words, accessing a formal **signal** parameter is the same as accessing an actual **signal**. For example, the signal assignments in the previous procedure example will change the value of the signals passed in through procedure calls.

Resolution Functions

A signal normally has a single source that is driving it. VHDL allows the designer to drive a signal with more than one source, if a *resolution function* is supplied to resolve the multiple sources into a single value for the signal. Resolution functions determine the value of a set of drivers to a signal. For example, in Figure 6.1(a) process i and process j both drive the same signal. Therefore, we need a resolution function to resolve the value of the driven signal.

For simulation, the resolution function can be any function consisting of any VHDL program. In a real circuit, the effect of the resolution function is to wire many signals together through a special gate that produces a single output. There are only limited ways of handling resolution functions in hardware such as wired-or gate, wired-and gate, and tri-state gate (Figure 6.1(b)).

A resolved signal is created if a signal declaration includes a resolution function or the declaration of the subtype for the signal contains a resolution function. An example for such a declaration is shown here:

```
signal NODE : WIRE_AND BIT ;
subtype RESOLVED_STD is WIRE_OR STD_ULOGIC ;
```

The first declaration makes signal NODE a resolved signal, where the WIRE_AND is the resolution function. Each time an event occurs to signal NODE, the WIRED_AND function is called and a value of BIT type will be returned. The

Subprograms and Packages

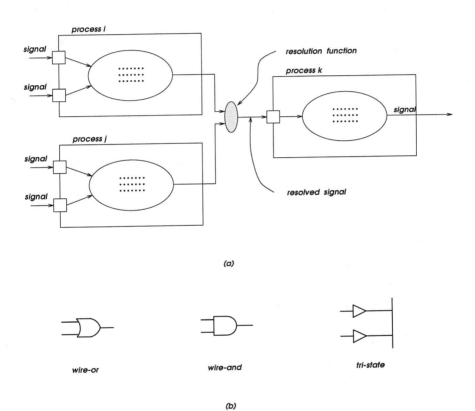

Figure 6.1 Behavior Model (a) Resolution function (b) Three hardware implementations.

second declaration defines a resolved subtype called RESOLVED_STD. Signals declared with RESOLVED_STD type are resolved signals.

The following shows a procedure to define and use a resolved signal:

1. define the signal's type (if necessary).
2. define the resolution function which takes input signals and returns a signal of this type.
3. declare a subtype of the signal type with an association with the resolution function.
4. declare and use resolved signals.

In the following example, Z is declared as a resolved type WIREOR_STD, where the resolution function of the data type is Wire_Or. In the architecture, there are two concurrent signal assignment statements. Both will assign value to signal Z. Z will get the value which is the "wire-or" of (I1 and I2) and (I3 xor I4).

```
architecture RESOLVED_ARCH of WOR is
    -- mebs wire_or
    function Wire_Or (DIN: in STD_ULOGIC_VECTOR)
        return STD_ULOGIC is
    begin
        -- mebs synthesis_off
        -- the mebs system will ignore this section
        return(DIN(0) or DIN(1)) ;
        -- mebs synthesis_on
    end Wire_Or ;
    subtype WIREOR_STD is  Wire_Or  STD_ULOGIC ;
    signal I1,I2,I3,I4  :  STD_ULOGIC ;
    signal Z            :  WIREOR_STD ;
begin
    Z <= I1 and I2 ;
    Z <= I3 xor I4 ;
end IMPLEMENT ;
```

Subprograms and Packages 95

The package STD_LOGIC_1164 of the IEEE library contains resolved data types, STD_LOGIC and STD_LOGIC_VECTOR. The resolved type STD_LOGIC is a tri-state resolved type. Multiple sources assigned to a signal of STD_LOGIC type will be resolved by a tri-state function.

6.2 PACKAGES

Data types, constants and subprograms can be declared inside entity declarations or inside architecture bodies. These declarations are visible in their associated architecture bodies. However, they are not visible to other entity declarations. To allow the same declarations to be visible by a number of design entities, VHDL provides packages. A package consists of two parts: a package declaration section and a package body.

The package declaration defines the interface for the package. The syntax of a package declaration is

> **package** *package_name* **is**
>   { *package_declarative_item* }
> **end** [*package_name*] ;

The *package_declarative_item* can be any of these:

- type declaration
- subtype declaration
- signal declaration
- constant declaration
- alias declaration
- component declaration
- subprogram declaration
- use clause (to include other packages)

Signal declarations in a package pose some problems in synthesis because a signal cannot be shared by two entities. A common solution to the problem is to make it as a "local" signal. In other words, if two entities both use the same signal from a package, each entity has its own copy of that signal.

The following shows an example of package declaration.

```
library IEEE ;
use IEEE.NUMERIC_BIT.all ;
package WATCH_PKG is
   subtype MONTH_TYPE is INTEGER range 0 to 12 ;
   subtype DAY_TYPE   is INTEGER range 0 to 31 ;
   subtype BCD4_TYPE  is UNSIGNED(3 downto 0) ;
   subtype BCD5_TYPE  is UNSIGNED(4 downto 0) ;
   constant BCD5_1   :  BCD5_TYPE := B"0_0001" ;
   constant BCD5_7   :  BCD5_TYPE := B"0_0111" ;
   function BCD_INC (L : in BCD4_TYPE) return BCD5_TYPE ;
end WATCH_PKG ;
```

The package body specifies the actual behavior of the package. A package body always has the same name as its corresponding package declaration, preceded by the reserved words **package body**. The information in a package body cannot be seen by designs or entities that use the package. The syntax of a package body is

> **package body** *package_name* **is**
> { *package_body_declarative_item* }
> **end** [*package_name*] ;

where *package_name* is the name of the associated package. A *package_body_declarative_item* is any of these:

- type declaration
- subtype declaration
- constant declaration

Subprograms and Packages 97

- subprogram body
- use clause

The following example shows a package body definition for package WATCH_PKG. All the declarations defined in a package declaration section are inherited into the corresponding package body.

```
package body WATCH_PKG is
    -- mebs module
    function BCD_INC (L : in BCD4_TYPE) return BCD5_TYPE is
        variable V, V1, V2 :  BCD5_TYPE ;
    begin
        V1 := L + BCD5_1 ;
        V2 := L + BCD5_7 ;
        case V2(4) IS
            when '0' => V := V1 ;
            when '1' => V := V2 ;
        end case ;
        return (V) ;
    end BCD_INC ;
end WATCH_PKG ;
```

Items declared inside a package declarations are not automatically visible to another library unit. A **use** clause preceding a unit will make items declared in a package declaration visible in the unit. For example, assume the previous package has been compiled into library MY_LIBRARY. To use the declarations in the package, we need to include them into the current library unit using the following statements.

```
library MY_LIBRARY ;
use MY_LIBRARY.WATCH_PKG.all;
```

IEEE standard committee defines two libraries for VHDL, namely STD and IEEE, each contains several packages.

- Library STD contains two packages: STANDARD and TEXTIO. The STANDARD package (1076-1987) defines some useful data types, such as INTEGER, BIT, BOOLEAN, *etc* as discussed in subsection 3.2.3. The

STANDARD package is automatically visible and no **use** clause is required. The TEXTIO package defines types and operations for communication with a standard programming environment. This TEXTIO package is used mostly for simulation and is not needed for synthesis.

- Library IEEE contains an IEEE Std 1164-1993 Standard Logic package (STD_LOGIC_1164), and Standard Synthesis packages, NUMERIC_BIT and NUMERIC_STD (pending for approval). These three packages contain some useful types and functions for VHDL simulation and synthesis.

6.2.1 IEEE Standard Logic Package

A problem with the STANDARD package is that it does not specify any multi-valued logic types for simulation purposes. The IEEE Standard Logic Package defines the following 9-valued logic.

```
type Std_ULogic is ('U','X','0','1','Z','W','L','H','-');
```

The meanings of the different type values of the nine values are given below.

'U'	Uninitialized
'X'	Forcing Unknown
'0'	Forcing Low
'1'	Forcing High
'Z'	High Impedance
'W'	Weak Unknown
'L'	Weak Low
'H'	Weak High
'-'	Don't Care

These nine values are meaningful for simulators. During simulation, a weak value on a node can always be overwritten by a forcing value. The high impedance can be overwritten by all other values. '0' and 'L' represent a logic level ground ("GND") and '1' and 'H' represent a logic level voltage supply ("VCC" or "VDD").

The 'U', 'W','X' and '−' are called *metalogical values*. The value 'U' represents the value of an object before it is explicitly assigned a value during simulation; the value 'X' represents forcing a node to be a value that the model is not able to distinguish between logic levels (unknown); the value 'W' represent forcing a weak value to a node; and the value '−' is called *don't care value*. In VHDL

Subprograms and Packages 99

simulation, they have a special interpretation when they are used in arithmetic or logic operations.

The value 'Z' is called the *high impedance value*. It represents the condition of a signal source when it makes no effective contribution to the resolved value of the signal.

Only the values '0', '1', 'X', '-', and 'Z' have a well defined meaning for a synthesis tool.

A signal of STD_ULOGIC type can have only one driver. If more than one driver is needed, a resolution function has to be defined. In the STD_LOGIC_1164 package, a subtype called STD_LOGIC is derived from STD_ULOGIC. STD_LOGIC is a resolved data type which comes with a predefined resolution function called RESOLVED. A signal of STD_LOGIC type can have more than one driver in a VHDL description.

The multiple value logic data type can be a vector. For example,

```
signal S1 : STD_ULOGIC_VECTOR(3 downto 0) ;
signal S2 : STD_LOGIC_VECTOR(7 downto 0) ;
```

There are some other useful functions defined in the package. For details of the definitions of the standard logic package, please refer Appendix B. The following summarizes some commonly used functions in the package.

- **to_bit(S, XMAP)**: converts a data object of STD_ULOGIC or STD_LOGIC type to BIT type. If the value of S is not '0' or '1', maps it into '0' or '1' according to the function specified in XMAP.

- **to_StdULogic(S)**: converts S of BIT type to a data object of STD_ULOGIC type.

- **rising_edge(S)**: detects the leading edges of signal S of type STD_ULOGIC.

- **falling_edge(S)**: detects the falling edges of signal S of type STD_ULOGIC.

To use the definitions and functions of the STD_LOGIC_1164 packages, the following statements must be included in a program.

```
library IEEE ;
use IEEE.STD_LOGIC_1164.all ;
```

6.2.2 IEEE Standard Synthesis Packages

Data objects of INTEGER type allow users to perform arithmetic computation, but do not provide an easy way to access part of the integer word. Data objects of BIT_VECTOR type allow users to access partial words, but do not allow users to perform arithmetic computation. For this reason, the type SIGNED and UNSIGNED are defined. A data object of SIGNED or UNSIGNED data type is treated as a bit vector but it can be used in arithmetic computation.

The Standard Synthesis packages have not yet become an IEEE standard (They are pending for approval). Standard synthesis packages include two VHDL packages: NUMERIC_BIT and NUMERIC_STD. The NUMERIC_BIT package is based on type BIT, while the NUMERIC_STD package is based on type STD_LOGIC. Both packages interpret the type UNSIGNED as an unsigned binary integer and the type SIGNED as a two's complement binary integer.

There are some other useful functions defined in the packages. For details of the definitions of the Standard Synthesis package, please refer Appendix B.

- **to_integer(arg)**: converts an object of unsigned or a signed type into an object of integer type.
- **to_signed(arg, n)**: converts an integer into a signed number of n bits.
- **to_unsigned(arg, n)**: converts an integer into an unsigned number of n bits.
- **resize(arg, n)**: changes the size of an unsigned or a signed number.
- **rising_edge(S)**: detects the leading edges of signal S of BIT data type.
- **falling_edge(S)**: detects the falling edges of signal S of BIT data type.
- arithmetic operations on unsigned or unsigned numbers such as **shift_left (arg, n)**, **shift_right (arg, n) rotate_left (arg, n)**, and **rotate_right (arg, n)**.

Both packages are compiled into the IEEE library. Since the numerical array types in both packages are defined using the same name, only one package can be used at a time.

To use the definitions and functions of the NUMERIC_BIT packages, the following statements have to be included in a program.

        library IEEE ;
        use IEEE.NUMERIC_BIT.all ;

To use the definitions and functions of the NUMERIC_STD packages, you have to include the following statements in the program:

        library IEEE ;
        use IEEE.STD_LOGIC.all ;
        use IEEE.NUMERIC_STD.all ;

6.3 SUMMARY

1. VHDL provides two kinds of subprograms: *procedures* and *functions*. A procedure is invoked as a statement and a function is invoked as an expression. A procedure can return none, one, or more than one argument, while a function always returns just one argument.

2. To make a set of type, constant, and subprogram declarations visible to a number of design entities, VHDL provides packages.

3. The STANDARD package in library STD is included in all VHDL source files by an implicit *use clause*.

4. The TEXTIO package in library STD defines types and operations for communication with a standard programming environment. It is used for simulation.

5. The IEEE Standard Logic Package in library IEEE defines package STD_LOGIC_1164 for operations of multiple value logic values. It is used to model signal wires which represent multiple values.

6. The IEEE Standard Synthesis Packages, NUMERIC_BIT and NUMERIC_STD, are used for synthesizing binary digital electronic circuits.

Exercises

1. A useful gate type in VHDL operator is EXCLUSIVE-NOR, which may be defined by the following equation:

$$z = \overline{x_1 \oplus x_2 \oplus ... \oplus x_n}$$

   Design a function for the EXCLUSIVE-NOR function with $n = 2$ and $n = 3$.

2. Write a function for converting integers between 0 to 255 to an eight-bit vector.

3. Write a function to check whether an eight-bit vector is odd parity.

4. Write a function to check whether a sixteen-bit vector is odd parity.

5. Write a function to convert a BCD number (4-bit) to a seven-bit vector for a seven segment display. The following figure shows the seven control signals to light up a seven segment display.

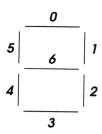

6. Write a procedure to perform an addition of two BCD numbers.

7. A subprogram can be expanded into the main program. Expand the following function into the architecture.

    ```
    entity INLINE is
    port ( IN1, IN2:  in INTEGER range 0 to 255;
           CTRL: in BIT ;
           VALUE1, VALUE2:  out INTEGER range 0 to 255) ;
    end INLINE ;
    architecture EXPAND of INLINE is
    subtype  BIT8 is INTEGER range 0 to 255;
    ```

Subprograms and Packages

```
      function ALU (L,R : BIT8 ; CT : BIT) return BIT8 is
         variable RESULT : BIT8 ;
      begin
         if (CT = '0') then
            RESULT := ( L + R ) mod 256 ;
         else
            RESULT := ( L - R ) mod 256 ;
         end if ;
         return ( RESULT ) ;
      end ALU ;
      begin
         VALUE1 <= ALU(IN1, IN2, CTRL);
         VALUE2 <= ALU("00001111", IN2, '1');
      end EXPAND ;
```

8. What is resolution function? Why do we need resolution functions?

9. Given the circuit in Fig. X6.1. There are two drivers for signal **Ou** and the resolution function for the signal is **wire-and**. Write a VHDL program for the circuit.

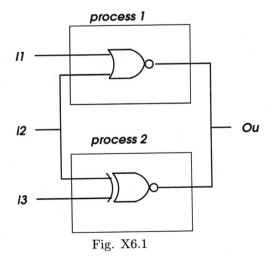

Fig. X6.1

10. If the resolution function is tri-state gate function, we can declare the output signal to be data type of STD_LOGIC. Write a VHDL program for the circuit in Fig. X6.1 assuming the resolution function is a tri-state gate function.

7

MODELING AT THE STRUCTURAL LEVEL

A digital system is usually represented as a hierarchical collection of components. Each component has a set of ports which communicate with the other components. A design can contain instances of lower-level designs, connected by signals to the lower-level designs' ports. In a VHDL description, a design hierarchy is introduced through component declarations and component instantiation statements.

While the basic unit of a behavior description is the process statement, the basic unit of a structural description is the component instantiation statement. Both the process statements and the component instantiation statements must be enclosed in an architecture body, which is a separate analyzable design unit. One of the important characteristics of VHDL is that it can model a design in mixed levels of abstraction; in other words, a description of an architecture may contain process statements and component instantiation statements.

7.1 COMPONENT DECLARATIONS

An architecture body can use other entities described separately and placed in the design libraries using component declaration and component instantiation statements. In a design description, each component declaration statement corresponds to an entity. In order to simulate or synthesize the current design, the entity and the architecture descriptions for all the components must be compiled in the design library.

The component declaration statement is similar to the entity specification statement in that it defines the component's interface. A component declaration is required to make a design entity usable within the current design. In other words, a component must be declared before is used (instantiated).

The syntax for a component declaration is

> **component** *component_name*
>   [**port** (*local_port_declarations*)]
> **end component** ;

where *component_name* is an *identifier* which represents the name of the component, and the syntax of *local_port_declaration* is the same as that defined for entity declaration. It defines the names, the attributes and the data types of the ports of the component. In the following, we will use a full adder to illustrate the concept of structure modeling. Fig. 7.1 (a) shows the interface for the full adder, and Fig. 7.1 (b) shows an implementation of the full adder. In this implementation, three types of components – OR2_gate, AND2_gate, and XOR_gate, are used to build the full adder circuit. The definitions of these three components are shown in the following:

```
component OR2_gate
    port (I0, I1 :  in STD_LOGIC; O : out STD_LOGIC);
end component;
component AND2_gate
    port (I0, I1 :  in STD_LOGIC; O : out STD_LOGIC);
end component;
component XOR_gate
    port (I0, I1 :  in STD_LOGIC; O : out STD_LOGIC);
end component;
```

7.2 COMPONENT INSTANTIATIONS

A component defined in an architecture may be instantiated using a component instantiation statement. At the point of instantiation, only the external view of the component (the names, types, and directions of its ports) is visible; signals internal to the component are not visible. The component instantiation statement instantiates and connects components to form a netlist of instances of a design.

Modeling at the Structural Level

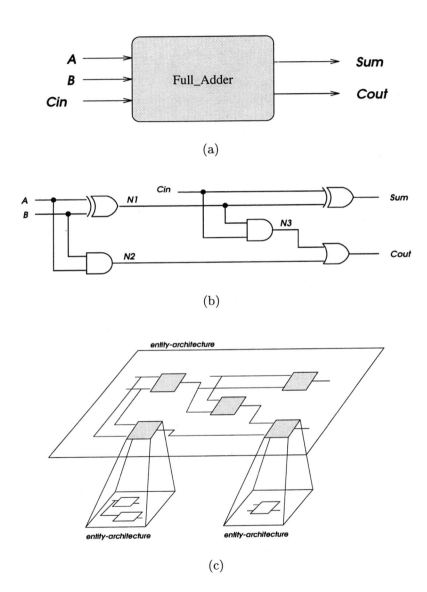

Figure 7.1 Full adder Design (a) Interface (b) Architecture (c) A Design Hierarchy.

The syntax of the component instantiation statement is

> *instantiation_label: component_name*
> **port map** (
>   [*local_port_name* ⇒] *expression*
>   { , [*local_port_name* ⇒] *expression* }
>) ;

where *instantiation_label* is the name of this instance of component type (*component_name*). A component instantiation statement must be preceded by an *instantiation_label*. These labels are referenced in component configurations which will be discussed later.

The *port map* phrase maps the component's ports onto signal nets. A port on a component declaration is called a *local* and a port on a component instance is called an *actual*. In a component instantiation statement, the port association list associates an *actual* with each *local*. The *actual* must be an object of class **signal**. Either positional or named association can be used in a component instantiation statement to specify port connections. With positional association, the expressions for the component ports are simply listed in the declared port order. With named association, the *port_name* ⇒ construct binds a local port to an actual signal. The following example shows two different types of port associations.

```
architecture INST of STRUCTURE is
   component XOR_gate
      port (I0, I1 :  in STD_LOGIC; O : out STD_LOGIC) ;
   end component ;
   component AND2_gate
      port( I0, I1 :  in  STD_LOGIC; O : out STD_LOGIC) ;
   end component ;
   signal A, B, Tmp1, Tmp2 :   STD_LOGIC ;
   -- use a default binding
begin
   U1: XOR_gate
      port map( A, B, Tmp1) ; --  Positional Association
   U2:  AND2_gate
      port map( I0 => A ,
```

Modeling at the Structural Level

```
                    I1 => B ,
                    O  => Tmp2) ; -- Named Association
    end INST ;
```

Since each component represents another design entity, a component instantiation in fact creates a level of hierarchy to the current design. Fig. 7.1 (c) shows the concept of design hierarchy in VHDL.

7.3 GENERATE STATEMENTS

The generate statement is a concurrent statement that has to be defined in an architecture. It is used to describe replicated structures. The syntax of a generate statement is

> *instantiation_label: generation_scheme* **generate**
> $\quad$ { *concurrent_statement* }
> **end generate** [*instantiation_label*] ;

There are two kinds of generation scheme: the **for** scheme and the **if** scheme. A *for-scheme* is used to describe a regular structure. It declares a generate parameter and a discrete range just as the for-scheme which defines a loop parameter and a discrete range in a sequential loop statement. The generate parameter needs not be declared. Its value may be read but cannot be assigned or passed outside a generate statement.

For example, we can use the generate statement to define a four-bit adder as follows. Fig. 7.2 (a) shows a schematic of a generated adder.

```
    architecture GEN_FOR of FULL_ADDER_4 is
        signal X, Y, Z : STD_LOGIC_VECTOR(3 downto 0) ;
        signal Cout    : STD_LOGIC ;
        signal TMP : STD_LOGIC_VECTOR(4 downto 0) ;
        component FULL_ADDER
            port (A, B, C : in STD_LOGIC;
                  S, Co :   out STD_LOGIC) ;
        end component ;
```

```
    begin
       TMP(0) <= '0' ;
       G :   for I in 0 to 3 generate
                FA : FULL_ADDER port map (
                            X(I), Y(I), TMP(I),
                            Z(I), TMP(I+1) ) ;
             end generate ;
       Cout <= TMP(4) ;
    end GEN_FOR ;
```

Many regular structures have some irregularities at the extreme. An *if-scheme* is designed to cope with these irregularities. Unlike the sequential if-statement, the if-generate cannot have **else** or **elsif** branches.

The following program uses the if-scheme to generate a four-bit full adder using half adder component and full adder component as shown in Fig. 7.2 (b).

```
    architecture GEN_FOR of FULL_ADDER_4 is
       signal X, Y, Z : STD_LOGIC_VECTOR(3 downto 0) ;
       signal Cout    : STD_LOGIC ;
       signal TMP : STD_LOGIC_VECTOR(4 downto 1) ;
       component HALF_ADDER
          port (A, B  : in STD_LOGIC;
                S, Co : out STD_LOGIC) ;
       end component ;
       component FULL_ADDER
          port (A, B, C : in STD_LOGIC;
                S, Co : out STD_LOGIC) ;
       end component ;
    begin
       G0:  for I in 0 to 3 generate
          G1:    if I = 0 generate
             HA : HALF_ADDER port map (
                         X(I), Y(I),
                         Z(I), TMP(I+1) ) ;
                 end generate ;
```

Modeling at the Structural Level

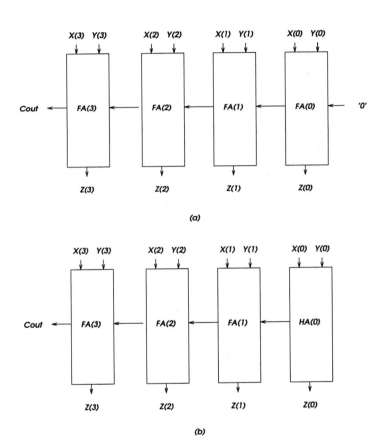

Figure 7.2 Generated Adder (a) for-scheme (b) if-scheme

```
            G2:    if I >= 1 and I <= 3 generate
                FA : FULL_ADDER port map (
                            X(I), Y(I), TMP(I),
                            Z(I), TMP(I+1) ) ;
                end generate ;
            end generate ;
            Cout <= TMP(4) ;
        end GEN_FOR ;
```

Any VHDL concurrent statement – process statement, concurrent signal assignment statements, concurrent procedure calls, block statement, concurrent assertion statement, and another generate statement – may be enclosed in a generate statement. Using nested generate statements, we can describe another kind of regularity. For example,

```
        Nested_Gen:  block
            L1 :    for I in 0 to 3 generate
                L2 :    for J in 0 to 3 generate
                    FA : CELL port map ( A(I), B(I), C(2*I+J), D(I+2*J) ) ;
                end generate ;
            end generate ;
        end block Nested_Gen ;
```

7.4 DEFAULT BINDINGS

The component declarations and component instantiations contain only the external view of the components used. It is not simulatable because there is no information in a program describing the functionalities of each component. We need a mechanism to link the component external view to the internal view. This is the concept of *configuration*. Let us use the full adder in Fig. 7.1 to illustrate the concept. The following example shows the structure description of the full adder.

```
        library IEEE ;
        use IEEE.STD_LOGIC_1164.all ;
        entity FULL_ADDER is
            port( A, B, Cin  :   in  STD_LOGIC ;
```

```
                Sum, Cout  :  out STD_LOGIC) ;
       end FULL_ADDER ;
       architecture IMP1 of FULL_ADDER is
          component XOR_gate
             port (I0, I1 :  in STD_LOGIC; O : out STD_LOGIC) ;
          end component ;
          component AND2_gate
             port( I0, I1 :  in  STD_LOGIC; O : out STD_LOGIC) ;
          end component ;
          component OR2_gate
             port( I0, I1 :  in  STD_LOGIC; O : out STD_LOGIC) ;
          end component ;
          signal N1, N2, N3 :  STD_LOGIC ;
          -- use a default binding
       begin
          U1 :    XOR_gate   port map (A, B, N1) ;
          U2 :    AND2_gate  port map (A, B, N2) ;
          U3 :    AND2_gate  port map (Cin, N1, N3) ;
          U4 :    XOR_gate   port map (Cin, N1, Sum) ;
          U5 :    OR2_gate   port map (N3, N2, Cout) ;
       end IMP1 ;
```

A structural description specifies what types of components will be used and how they are connected, but does not provide functionality description for each component. To simulate this program, we need to describe the behavior of each component. The following shows the entity and architecture specifications of the components used in the full adder design.

```
       library IEEE ;
       use IEEE.STD_LOGIC_1164.all ;
       entity AND2_gate is
           port (I0, I1 :  in STD_LOGIC; O : out STD_LOGIC) ;
       end AND2_gate ;
       architecture BHV of AND2_gate is
       begin
            O <= I0 and I1 ;
       end BHV ;
```

```
library IEEE ;
use IEEE.STD_LOGIC_1164.all ;
entity OR2_gate is
    port (I0, I1 :  in STD_LOGIC; O : out STD_LOGIC) ;
end OR2_gate ;
architecture BHV of OR2_gate is
begin
    O <= I0 or I1 ;
end BHV ;
library IEEE ;
use IEEE.STD_LOGIC_1164.all ;
entity XOR_gate is
    port (I0, I1 :  in STD_LOGIC; O : out STD_LOGIC) ;
end XOR_gate ;
architecture BHV of XOR_gate is
begin
    O <= I0 xor I1 ;
end BHV ;
```

When all the entities and architectures of the components have been compiled into the working library, we can proceed to compile the FULL_ADDER design. When we compile the FULL_ADDER design, the compiler will match each component name with the entity name in the working library. If each component can find an entity which matches the names, the compiler will build a simulation object for the FULL_ADDER design. For example, the three types of components in architecture IMP1 are XOR_gate, AND2_gate, and OR2_gate. If there are entities with the same names in the library WORK, the simulator will bind their last compiled architecture (in case an entity has more than one architecture) to the corresponding component. Since a designer does not have to explicitly specify the binding between the instantiated component and the entity-architecture pair in the design library, we call it *default binding*.

In case there are multiple architectures for the entity being bound to a component of a design. The compiler will use the last compiled architecture of an entity to build the executable for the simulator. Using the last compiled architecture for an entity to build the simulator will work fine in a component entity with only one architecture. When there is more than one architecture for an entity, it can become confusing as to which architecture was compiled

Modeling at the Structural Level

last. A better method is to specify exactly which architecture is to be used for each entity. This can be done by *configuration specification* and *configuration declaration* which will be discussed in the next two sections.

7.5 CONFIGURATION SPECIFICATIONS

A single entity may contain several architectures. One architecture might be an algorithmic model, one architecture might be at register transfer level, while another architecture might be a structural model for the entity. At some point in a design process a designer may wish to specify, for each component instance, exactly which entity declaration in which design library and which architecture of that design entity is to be selected. The *configuration specification* is the construct that allows the designer to specify the selection of entity declaration and architecture body for each component instance inside the architecture body of the design.

The syntax of component configuration specification is

> **for** *instantiation_list: component_name* **use** *binding_indication* ;

where the *instantiation_list* identifies the instances of the component (specified by *component_name*) to be configured by the configuration specification. The instances can be specified by stating the *instantiation_labels* or using keywords **others** or **all**. The keyword **others** refers to the rest of instances of component *component_name* which have not been configured. The keyword **all** refers to all the instances of component *component_name*.

The *binding indication* specifies the mapping between component declaration and entity declaration. It specifies the library, entity and an optional architecture that the instances will be bound. A general form for the *binding_indication* is

> **entity** *library_name.entity_name* [(*architecture_name*)] ;

If there is only one architecture in an entity, the architecture name in the *binding_indication* can be omitted. The following shows the usage of configuration specification for the full adder example.

```
library IEEE ;
use IEEE.STD_LOGIC_1164.all ;
entity FULL_ADDER is
   port( A, B, Cin  :  in  STD_LOGIC ;
         Sum, Cout  :  out STD_LOGIC) ;
end FULL_ADDER ;
architecture IMP1 of FULL_ADDER is
   component XOR_g
      port (I0, I1 : in STD_LOGIC; O : out STD_LOGIC) ;
   end component ;
   component AND2_g
      port( I0, I1 : in STD_LOGIC; O : out STD_LOGIC) ;
   end component ;
   component OR2_g
      port( I0, I1 : in STD_LOGIC; O : out STD_LOGIC) ;
   end component ;
   signal N1, N2, N3 : STD_LOGIC ;
   for  U1:    XOR_g     use entity work.XOR_gate(BHV) ;
   for  others: XOR_g    use entity work.XOR_gate(BHV) ;
   for  all:   AND2_g    use entity work.AND2_gate ;
   for  U5:    OR2_g     use entity work.OR2_gate ;
begin
   U1 :   XOR_g  port map (A, B, N1) ;
   U2 :   AND2_g port map (A, B, N2) ;
   U3 :   AND2_g port map (Cin, N1, N3) ;
   U4 :   XOR_g  port map (Cin, N1, Sum) ;
   U5 :   OR2_g  port map (N3, N2, Cout) ;
end IMP1 ;
```

7.6 CONFIGURATION DECLARATIONS

Binding a component instantiation to an actual entity-architecture pair does not have to be done in the architecture that uses this component. The binding can be deferred and accomplished later using *configuration declaration* statements. By delay binding the real object, it is possible to describe a generic design in advance and bind the real objects at a later stage. Also, a generic design can be configured into several different implementations which is useful in documenting different versions of a design during the design process.

A configuration is a design unit that can be compiled separately and stored in a library. It saves time because a user does not have to recompile the entire design if he needs to substitute only a few components.

A configuration declaration begins by giving the configuration a name, and associating it with an entity; in other words, each configuration declaration defines a configuration for a particular entity. For example, a configuration declaration for the full adder begins with the following line:

```
configuration CONFIG1 of FULL_ADDER is
```

After the initial declarations, a set of **for ... use** statements provide the configurations for the architectures, blocks, or components in the entity. After the configuration is complete, all the components in the entity must be bound (either explicitly or implicitly).

For example, the following declares a configuration called CONFIG1 for architecture IMP1 of entity FULL_ADDER.

```
configuration CONFIG1 of FULL_ADDER is
   for IMP1
   - - default binding used
   end for ;
end CONFIG1 ;
```

The configuration declaration begins with the key word **configuration** and is followed by the name of the configuration called CONFIG1. The keyword **of** precedes the name of the entity being configured. The second line of the configuration starts the block configuration section. The keyword **for** is followed by a name of the architecture to be configured or the name of the block of the architecture to be configured. Any component or block information will then exist between the **for** clause and the matching **end for**. In this example, there are no blocks or components to configure, therefore, the block configuration area is empty, and default binding will be used. Therefore, the *CONFIG1* binds architecture *IMP1* with entity *FULL_ADDER* to form a simulatable object.

We can bind the instances within an architecture using a **for** clause and a matching **end for** clause in a configuration declaration. Let's use the full adder as an example to illustrate the component configurations. The implementation of the full adder design consists of three types of components: XOR_g, AND2_g, and OR2_g. Assume entities XOR_gate, AND2_gate, and OR2_gate have been compiled into library WORK. The following configuration specifies for each component instance in the architecture IMP1 a mapping to an entity-architecture pair in the library. Each component being configured has a **for** clause to begin the configuration, and an **end for** clause to end the configuration. Each component can be specified with the component instantiation label directly, or with an **all** or **others** clause.

```
configuration CONFIG1 of FULL_ADDER is
   for IMP1
      for U1:    XOR_g  use entity work.XOR_gate(BHV) ;
      end for ;
      for others:  XOR_g  use entity work.XOR_gate(BHV) ;
      end for ;
      for   all:  AND2_g use entity work.AND2_gate ;
      end for ;
      for    U5:  OR2_g  use entity work.OR2_gate(BHV) ;
      end for ;
   end for ;
end CONFIG1 ;
```

Alternatively, an entity-architecture pair can be configured as a library unit and used by a high level configuration. For example, assume the following XOR_gate entity has been configured as a library unit called XOR_config.

```
library IEEE ;
use IEEE.STD_LOGIC_1164.all ;
entity XOR_gate is
   port (I0, I1 :  in STD_LOGIC; O : out STD_LOGIC) ;
end XOR_gate ;
architecture BHV of XOR_gate is
```

```
begin
   O <= IO xor I1 ;
end BHV ;
configuration XOR_config of XOR_gate is
   for BHV
   end for ;
end XOR_config ;
```

Then, the **use configuration** clause can specify the configuration to use for this instance of the component. For the full adder example, the configuration specification for component U1 can use the configuration XOR_config instead of an entity-architecture pair from the working library. In order for CONFIG1 to compile, configuration XOR_config must be compiled into library WORK.

```
configuration CONFIG1 of FULL_ADDER is
   for IMP1
      for     U1:  XOR_g use  configuration work.XOR_config ;
      end for ;
      for others:  XOR_g use  configuration work.XOR_config ;
      end for ;
      for    all:  AND2_g use entity work.AND2_gate;
      end for ;
      for     U5:  OR2_g use  entity work.OR2_gate;
      end for ;
   end for ;
end CONFIG1 ;
```

Another use of configurations is to provide remapping of component ports. When the port names for an entity being configured do not match the component port name, we can use the port map clause to match the names. The configuration port map clause looks like the component instantiation port map clause used in an architecture. For example, if the entity-architecture pair of XOR_gate has input ports x and y, and an output port z. Then, we can match the name by adding a port remapping clause in the configuration statement.

```
configuration CONFIG1 of FULL_ADDER is
   for IMP1
      for     U1:  XOR_g use  configuration work.XOR_config ;
```

```
                        port map( x=>I0, y=>I1, z=>O) ;
      end for ;
      for others:   XOR_g use   configuration work.XOR_config ;
                        port map( x=>I0, y=>I1, z=>O) ;
      end for ;
      for    all:   AND2_g use configuration work.AND2_config ;
      end for ;
      for    U5:    OR2_g use  configuration work.OR2_config ;
      end for ;
   end for ;
end CONFIG1 ;
```

7.7 MODELING A TEST BENCH

To test a compiled VHDL design, we can provide stimuli interactively through the command window of a simulator. Alternatively, we can write a VHDL program (a VHDL test bench) to test a compiled VHDL design. A test bench does not have external ports. It contains a component representing the circuit under test, and waveform generators which produce waveforms to the inputs of the component under test. Fig. 7.3 illustrates a test bench design for the full adder in Fig. 7.1. The following shows a very simple VHDL test bench for the full adder circuit.

```
library IEEE ;
use IEEE.STD_LOGIC_1164.all ;
entity ADDER_Testbench is
end ADDER_Testbench ;
architecture TEST1 of ADDER_Testbench is
   component F_ADD
      port( A, B, Cin  : in  STD_LOGIC ;
            Sum, Cout  : out STD_LOGIC) ;
   end component;
   for U1:    F_ADD use entity WORK.FULL_ADDER(IMP1) ;
   signal A, B, Ci, Co, S : STD_LOGIC ;
begin
```

```
        U1:   F_ADD port map( A => A , B => B , Cin => Ci ,
                         Sum => S , Cout => Co) ;
        U2:    A    <=   '0' ,
                         '1' after  50 ns ,
                         '0' after 100 ns ,
                         '1' after 150 ns ,
                         '0' after 200 ns ,
                         '1' after 250 ns ,
                         '0' after 300 ns ,
                         '1' after 350 ns ;
        U3:    B    <=   '0' ,
                         '0' after  50 ns ,
                         '1' after 100 ns ,
                         '1' after 150 ns ,
                         '0' after 200 ns ,
                         '0' after 250 ns ,
                         '1' after 300 ns ,
                         '1' after 350 ns ;
        U4:   Ci    <=   '0' ,
                         '0' after  50 ns ,
                         '0' after 100 ns ,
                         '0' after 150 ns ,
                         '1' after 200 ns ,
                         '1' after 250 ns ,
                         '1' after 300 ns ,
                         '1' after 350 ns ;
   end TEST1 ;
```

The TEST1 architecture is the top unit of the VHDL hierarchy of the test bench. It includes the full adder entity as its component. A configuration specification binds component F_ADD to entity FULL_ADDER and architecture IMP1 in library WORK. Three signal assignments with timing specification are used to provide input waveforms to the component. Simulation begins at 0 ns, the inputs change every 50 ns, and simulation ends at at 350 ns. During the time period, the output signals of the full adder will change its value immediately after the inputs change values.

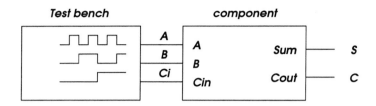

Figure 7.3 A test bench for the full adder

Configuration declaration can also be used to model a test bench. Using configuration declaration, binding a component instantiation to the actual component does not have to be done in the architecture that uses this component. This binding can be deferred until later and accomplished by a configuration declaration. This way, a single test bench can be used to test various versions of a design.

For example, in chapter 12 we will introduce a single pulser design with four architectures representing four different levels of abstraction. The function of a single pulser is very simple. When the input becomes '1', the circuit will generate an output pulse which lasts for one clock cycle. We can use a generic test bench to verify the correctness of the four designs. In the following test bench, we declare a component *pulser* to represent the single pulser design. The goal of the test bench is to generate waveforms to the input pins of the unit under test and get the values from the output pins.

```
use STD.TEXTIO.all ;
entity PULSE_TEST_BENCH is
end PULSE_TEST_BENCH ;
architecture SIM of PULSE_TEST_BENCH is
   component PULSER
     port ( CLK              : in  BIT ;
            DATA_IN          : in  BIT ;
            DATA_OUT         : out BIT ) ;
   end component ;
   file FILEIN      : TEXT IS IN  "pulser.input" ;
   file FILEOUT     : TEXT IS OUT "pulser.output" ;
   signal CLK       : BIT:= '1' ;
```

```
      signal DATA_IN  : BIT ;
      signal DATA_OUT : BIT ;
   begin
      TEST: PULSER port map (CLK, DATA_IN, DATA_OUT) ;
      CLK <= not CLK after 100 ns ;
      process
         variable LINEIN, LINEOUT : line ;
         variable TEMP    : BIT ;
        begin
         loop
            readline(FILEIN, LINEIN) ;
            exit when endfile(FILEIN) ;
            read(LINEIN, TEMP) ;
            DATA_IN <= TEMP ;
            wait until CLK'event and CLK='1' ;
            write(LINEOUT, now, left, 7) ;
            write(LINEOUT, DATA_IN, left, 7) ;
            write(LINEOUT, DATA_OUT, left, 7) ;
            writeline(FILEOUT, LINEOUT) ;
         end loop ;
      end process ;
   end SIM ;
```

In this test bench, the test data is read from file FILEIN (physical name "pulser.input") and the output is written into file FILEOUT (physical name "pulser.output"). The CLK signal to the single pulser has an initial value of '1' and it toggled every 100 nanoseconds. A process is written to read the test data from file FILEIN and the simulation result is written into file FILEOUT. The first four line in the loop means to read a data from FILEIN and apply it to port DATA_IN. Then, after the CLK leading edge, the results are written into file FILEOUT. Note **now** is a function defined in the IEEE.TEXTIO package which reports the simulation time. The parameter *left* means the output will be left justified and 7 represents the size of the output.

We defer the binding of actual entity design to the component unit under test in the test bench. The actual binding is done using a configuration declaration. Let entity **PULSE** and architecture **ALG** have been compiled into library

WORK. The following configuration declaration binds instance TEST in the test bench to the architecture ALG and entity PULSE.

```
configuration ALG_SIMULATE of PULSE_TEST_BENCH is
  for SIM
    for TEST: PULSER use entity WORK.PULSE(ALG) ;
    end for ;
  end for ;
end PULSE_TEST_BENCH ;
```

Similarly, if we want to simulate the other three architectures, we can simply write a configuration declaration for each architecture such as the following and have the design description ready. This way, we can save a lot of time to recompile the architectures.

```
configuration FSMD_SIMULATE of PULSE_TEST_BENCH is
  for SIM
    for TEST: PULSER use entity WORK.PULSE(FSMD) ;
    end for ;
  end for ;
end PULSE_TEST_BENCH ;
configuration RTL_SIMULATE of PULSE_TEST_BENCH is
  for SIM
    for TEST: PULSER use entity WORK.PULSE(RTL) ;
    end for ;
  end for ;
end PULSE_TEST_BENCH ;
configuration GATE_SIMULATE of PULSE_TEST_BENCH is
  for SIM
    for TEST: PULSER use entity WORK.PULSE(GATE) ;
    end for ;
  end for ;
end PULSE_TEST_BENCH ;
```

Modeling at the Structural Level

7.8 SUMMARY

1. In a VHDL program, entities and architectures cannot be nested. A design hierarchy is introduced through component declarations and component instantiation statements.

2. In a design, each component declaration statement in an architecture body corresponds to an entity.

3. A component defined in an architecture may be instantiated using a component instantiation statement.

4. At the point of instantiation, only the external view of the child component (the names, types, and directions of its ports) is visible; signals internal to the component are not visible.

5. The component declarations and component instantiations contain only the external view of the components used. It is necessary to bind each instance to an entity-architecture pair in order to make the program simulatable.

6. An instance of a component can be bound to an entity-architecture pair using *default binding, configuration specification,* or *configuration declaration.*

7. A VHDL test bench is a VHDL program which is used to test a compiled VHDL design.

Exercises

1. The following shows a 1-bit slice parallel loadable right shift register. Write a gate level description of the shift register.

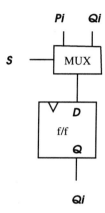

2. Use the 1-bit loadable right shift register in the previous exercise to build a 4-bit, right shift, parallel loadable shift register. Draw the schematic of the shift register and describe it in VHDL.

3. Use the full adder in Fig. 7.1 as a component, build an 8-bit adder and describe it in VHDL.

4. Make the adder in the previous example as a parameterized module using a generic constant.

5. A T-type (toggle) flip-flop has a single data input denoted T and is characterized by the fact that T = 1 causes the flip-flop to change state, while T= 0 retains the current state.

    (a) Give the state table and the characteristic equation defining the behavior of a (clocked) T-flip-flop.

    (b) Show how to convert a D flip-flop to a T flip-flop by adding some logic to it.

    (c) Describe the structural design in VHDL.

6. Carry out a gate level design of the following synchronous modulo-4 up-down counter. It has four input lines: a count enable line COUNT; an up-down select line DOWN; a clock line CLOCK; and a reset line CLEAR.

There are two output lines $Z = (z_0, z_1)$. Use only NAND gates and D flip-flops in your circuit, and attempt to minimize the total number of logic gates.

7. Explain the following terms:
   (a) default binding
   (b) configuration specification
   (c) configuration declaration
   (d) test bench

8. Derive the truth table for an 8-bit priority encoder. Minimize the encoder using minimal number of AND or OR gates. Write a structural VHDL for the design.

9. Design a 16-bit priority encoder using two copies of an 8-bit priority encoder designed in the previous problem. Additional gates may be used if needed. Use configuration specification to refer the instances of the 8-bit priority encoder. Write a structural VHDL for the design.

10. Consider the task of designing a two-level circuit that increments a number N represented in 4-bit BCD form, producing (also in BCD form) the result $N + 1$. When $N = 9$, $N + 1$ is taken to be zero, so that the summation is modulo-10. Assume you have only NAND gates (2-input and 3-input) available, design and minimize your circuit using minimum number of NAND gates. Describe the structural design in VHDL.

11. Consider the 1-digit BCD incrementer. If we are going to modify the previous 1-digit decimal incrementer by adding a fifth output signal that is normally 0, but becomes 1 if the input does not represent a BCD number. This bit thus serves the presence of erroneous input combinations. Design a minimal all-NOR (2-input or 3-input) implementation of the modified incrementer. Describe the circuit in VHDL and write a test bench for the design.

12. Use two 4-bit binary adders and the necessary AND/OR correction logic to design a single-digit BCD adder that can be bit-sliced to produce an n-bit BCD adder. The algorithm to carry out a BCD arithmetic is described as follows:

> Add two BCD numbers by using binary addition. If their sum is 1001 or less, the sum is valid and no correction is made. If their sum is greater than 1001, add 0110 to the sum to give the correct BCD result, and send the carry to the next most significant digit.

Derive a structural description for the circuit and write a test bench for it.

8
MODELING AT THE RT LEVEL

A Register Transfer Level (RTL) design consists of a set of registers connected by combinational logic as shown in Fig. 8.1. The registers are shown as rectangular boxes connected to the clock signal and the combinational logic is represented by the "cloud" object. This chapter explains the relationship between the RTL constructs in VHDL and the logic which is synthesized. It focuses on code styles that will give the best performance for an RTL synthesis tool. An RTL synthesis tool produces registered and combinational logic at the RTL level.

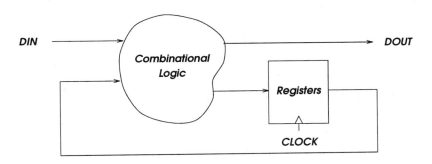

Figure 8.1 A representation of an RTL design.

8.1 COMBINATIONAL LOGIC

A process without **if** signal leading edge (or falling edge) statements or **wait** signal'event statements is called a *combinational process*. A combinational process is usually synthesized as a combinational circuit (with an exception on latch inferences which will be discussed later). The following example describes a combinational logic which generates the carry out signal for inputs A, B and Cin.

```
signal A, B, Cin, Cout : BIT ;
  . . .
process(A, B, Cin)
begin
    Cout <= (A and B) OR ((A or B) and Cin) ;
end;
```

In this example, all the input signals must be listed in the sensitivity list. It indicates that the process will be executed whenever the signals of the sensitivity list change. To describe a combinational circuit, the variables or signals in the process must not have initial values because a combinational logic does not have memory to hold a value. Furthermore, a signal or a variable must be assigned a value before being referenced.

All the sequential statements except **wait** statements, **loop** statements and **if** signal leading edge (or falling edge) statements can be used to describe a combination logic. The arithmetic operators (such as +, -, *, *etc*), relational operators (such as <, >, =, *etc*), and logic operators (such as **and**, **or**, **not**, *etc*) can be used in an expression.

A synthesis tool may perform resource sharing if there are mutually exclusive operations in a description. For example, the two addition operations in the following program are mutually exclusive operations (will not be executed simultaneously). We can assign them to two different function units or to the same function unit depending on the resource constraints. If the resource constraint is one adder, then the output circuit will be the one shown in Fig. 8.2(a). However, if the resource constraints are two adders, the resulting circuit will be that of Fig. 8.2(b).

```
process (A,B,Sel)
begin
    if (Sel = '1') then
```

Modeling at the RT Level

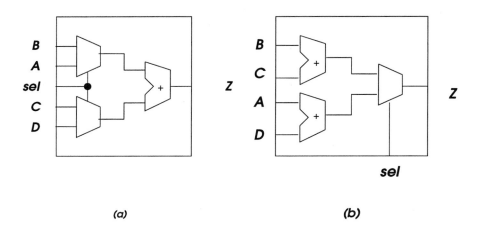

Figure 8.2 Two different synthesized results.

```
        Z <= B + C ;
    else
        Z <= A + D ;
    end if ;
end process ;
```

Concurrent signal assignments can be used as short hand notations for some combinational processes. For example, the above carry generation process can be rewritten as a concurrent signal assignment.

```
architecture Data_Flow of Full_Adder is
signal A, B, Cin, Cout :  BIT ;
 .  .  .
Cout <= (A and B) OR ((A or B) and Cin) ;
end Data_Flow ;
```

8.2 LATCHES

Flip-flops and latches are two commonly used one-bit memory devices. A flip-flop is an edge-triggered memory device. A latch is a level-sensitive memory device. The following example describes a level-sensitive latch:

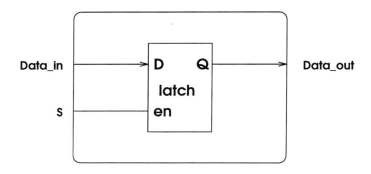

Figure 8.3 A simple latch.

```
-- latch inference
signal S, Data_In, Data_Out:  BIT ;
. . .
process (S, Data_In)
begin
   if (S = '1') then
      Data_Out <= Data_In ;
   end if ;
end process ;
```

The sensitivity list includes signals S and Data_In which are required for a latch inference. It indicates whenever the signals S and Data_In change, the process will be executed. Since the assignment to the signal Data_Out is hidden in a conditional clause, Data_Out will not change value (will preserve its old value) if S is '0'. If S is '1', Data_Out is updated with the value of Data_In, whenever the signals in the sensitivity list change value. This is the behavior of a level-sensitive latch (Fig. 8.3).

In general, latches are synthesized from incompletely specified conditional expressions in a combinational description. Any signal or variable that is not driven under all conditions becomes a latched element. Latch inferences occur normally with **if** statements or **case** statements. The above program infers a latch because the **if** statement is incompletely specified.

Modeling at the RT Level

To avoid having a latch inferred, assign a value to the signal under all conditions. For example, by adding an extra statement into the previous program as shown in the following example, the **if** statement becomes fully specified. Therefore, the program will be synthesized as an AND gate.

```
-- AND gate
signal S, Data_In, Data_Out:  BIT ;
 . . .
process (S, Data_In)
begin
   if (S = '1') then
      Data_Out <= Data_In ;
   else
      Data_Out <= '0' ;
   end if ;
end process ;
```

We can specify a latch with an asynchronous reset or an asynchronous preset. An asynchronous reset (or preset) will change the output of a latch to 0 (or 1) immediately. The following shows a latch which will be reset to '0' when the asynchronous input signal RST becomes '1'. The RST is active high in this example. We can change it to active low by changing the condition to "(RST = '0')". Fig. 8.4 shows a circuit diagram corresponding to the specification.

```
signal S, RST, Data_In, Data_Out:  BIT ;
 . . .
process (S, RST, Data_In)
begin
   if (RST = '1') then
     Data_Out <= '0' ;
   elsif (S = '1') then
     Data_Out <= Data_In ;
   end if ;
end process ;
```

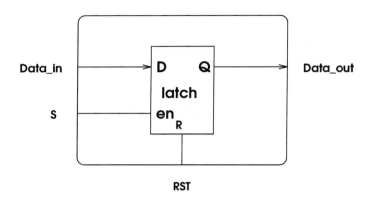

Figure 8.4 A latch with asynchronous reset.

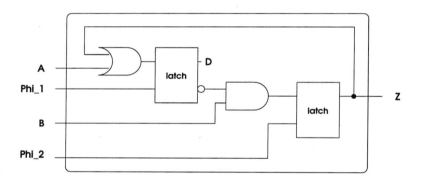

Figure 8.5 A digital design using two-phase clock.

8.3 DESIGNS WITH TWO PHASE CLOCKS

A two-phase clock control circuit can be described using latch inferences. The basic concept of describing a two-phase clock design is to partition the circuit into two processes. One process describes the first level combinational logic and latches, and the other process describes the second level combinational logic and latches. The following example shows a VHDL program which describes a two-phase clock controlled circuit of Fig. 8.5.

```
entity TwoPhase is
   port ( A, B            : in BIT ;
          Phi_1,Phi_2      : in BIT ;
          Z                : buffer BIT );
end TwoPhase ;
architecture Implement of TwoPhase is
   signal D : BIT ;
begin
   process (A, Z, Phi_1)
   begin
      if (Phi_1 = '1') then
         D <= A or Z ;
      end if ;
   end process ;
   process (B, D, Phi_2)
   begin
      if (Phi_2 = '1') then
         Z <= B and not D ;
      end if ;
   end process ;
end Implement ;
```

8.4 FLIP-FLOPS

A process with **if** signal leading edge (or falling edge) statements or **wait** signal'event statements is called a *clocked process*. An edge triggered flip-flop will be generated from a VHDL description if a signal assignment (or possibly variable assignment) is executed on the leading (or on the falling) edge of another signal. By detecting clock edges, the synthesis tool can locate where to insert flip-flops so that the design that is built behaves similar to what the simulation predicts.

The **event** attribute on a signal is the most commonly used edge-detecting mechanism. It operates on a signal and returns a boolean value. The result is true if the signal shows a change in value. The attribute **stable** is the boolean inversion of the **event** attribute.

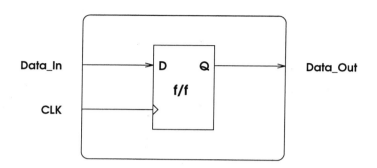

Figure 8.6 A D-type flip-flop.

The following example shows a simple VHDL program which describes a flip-flop. The process contains an **if** signal leading edge statement. During the leading edges of the signal CLK, the content of Data_In is assigned to Data_Out. This is exactly the behavior of a D flip-flop as shown in Fig. 8.6.

```
signal CLK, Data_In, Data_Out:  BIT ;
  . . .
process ( CLK )
begin
   if (CLK'event and CLK = '1') then
      Data_Out <= Data_In ;
   end if ;
end process ;
```

Variables can also generate flip-flops. Since the variable is defined in the process itself, and its value will never leaves the process, the only time a variable generates a flip-flop is when the variable is used before it is assigned in a clocked process. For example, the following code segments generate two flip-flops.

```
signal CLK, Data_In, Data_Out:  BIT ;
  . . .
process ( CLK )
variable TMP ;
begin
   if (CLK'event and CLK = '1') then
      Data_Out <= TMP ;
```

```
        TMP := Data_In ;
      end if ;
   end process ;
```

In this case, variable TMP is used before it is assigned. Since the value of TMP is the one that is assigned in the last run (one clock earlier) of the process evaluation, we need a flip-flop to keep its value. If we reverse the two assignments in the process, the TMP variable will become a wire. Therefore, only one flip-flop will be generated.

8.5 SYNCHRONOUS SETS AND RESETS

Synchronous inputs set (preset) or reset (clear) the output of flip-flops when they are asserted. The assignments will only take effect while the clock edge is active; at all other times, changes on these inputs are not noticed by the memory element.

All the signal assignments inside an "**if** signal leading (or falling) edge expression then" clause cause the values to be assigned to flip-flops along the clock edge. A flip-flop with synchronous reset can be modeled by detecting whether a synchronous input has been set on the leading edge (or on the falling edge) of a clock. For example, the following program corresponds to the circuit in Fig. 8.7(a).

```
   signal CLK, Data_In, Data_Out, S_RST: BIT ;
   . . .
   process ( CLK )
   begin
      if (CLK'event and CLK = '1') then
         if (S_RST = '1') then
            Data_Out <= '0' ;
         else
            Data_Out <= Data_In ;
         end if ;
      end if ;
   end process ;
```

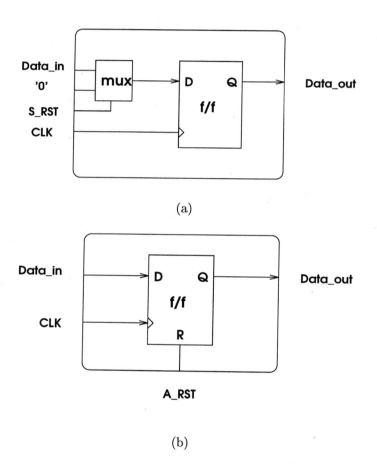

Figure 8.7 Reset (a) A D-type flip-flop with synchronous reset. (b) A D-type flip-flop with asynchronous reset.

Modeling at the RT Level 139

Note that it does not matter whether or not signals S_RST and Data_In are on the sensitivity list of the process because their changes do not result in any action inside the "**if** signal'edge **then**" clause.

8.6 ASYNCHRONOUS SETS AND RESETS

In a number of instances, D-type flip-flops are required to have an asynchronous set or reset inputs. Asynchronous inputs set (preset) or reset (clear) the output of flip-flops whenever they are asserted independent of the clock. The following VHDL program shows a VHDL program which describes a D-type flip-flop with an asynchronous reset input.

```
signal CLK, A_RST, Data_In, Data_Out: BIT ;
...
process (CLK, A_RST)
begin
   if (A_RST = '0') then
      Data_Out  <= '0' ;
   elsif (CLK'event and CLK = '1') then
      Data_Out  <= Data_In ;
   end if ;
end process ;
```

The entity now has an extra input, the A_RST port, which will be used to reset the D flip-flop asynchronously. The signals CLK and A_RST must be in the sensitivity list. Whenever an event occurs on either signal CLK or A_RST, the statements inside the process will be executed. Signal A_RST is first tested to see if it is '1' (active high asynchronous reset). If so, the output of the flip-flop is reset to '0'. Otherwise, the value of Data_In is assigned to the signal Data_Out at the leading edge of the signal CLK. Fig. 8.7(b) shows the schematic for such a flip-flop.

It is possible to describe a flip-flop with more than one asynchronous inputs. The following describes such a circuit.

```
signal CLK, RST, PRST, EN: BIT ;
signal Data_In, Data_Out: BIT ;
process (CLK, PRST, RST, EN)
```

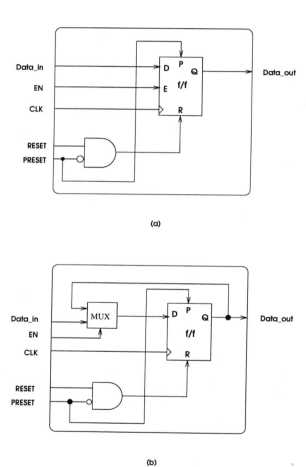

Figure 8.8 Two implementations of an enabled D-type flip-flop.

Modeling at the RT Level 141

```
begin
   if (PRST = '1') then
      Data_Out <= '1' ;
   elsif (RST = '1') then
      Data_Out <= '0' ;
   elsif (CLK'EVENT and CLK = '1') then
      if (EN = '1') then
         Data_Out <= Data_In ;
      end if ;
   end if ;
end process ;
```

This circuit performs a prioritized testing of the PRST signal with respect to the RST signal. The PRST signal will be tested before the RST signal. If the PRST signal is active, the flip-flop will be preset to '1' regardless of the state of the RST input. If the PRST signal is not active, the flip-flop will be reset to '0' when RST input is asserted. Otherwise, the Data_In is assigned to Data_Out synchronously when EN signal is asserted. Fig. 8.8 (a) shows the schematic of the circuit.

In the previous example, there is an enable input to the flip-flop. If there is a flip-flop with enable input available, a synthesizer may build a circuit in two ways. If there are flip-flips with enable input in the technology library, it will use it (Fig. 8.8 (a)). If there is no such a flip-flop, the synthesizer may use a multiplexer to build the circuit. Fig. 8.8 (b) shows such a circuit.

8.7 VHDL TEMPLATES FOR RTL CIRCUITS

Based on the previous discussion on the modeling of combinational logic and flip-flops, we can divide the statements of an RTL process into several *synchronous sections* and *combinational sections*. A *synchronous section* describes a sub-circuit whose behavior will be evaluated only on the signal's edges, and a *combinational section* describes a sub-circuit whose behavior will be evaluated whenever there is a change on the signals of the sensitivity list. All the signals referenced in a combinational section must be listed in the sensitivity list.

A signal assignment statement within a synchronous section infers an assignment of a value to a register. Fig. 8.9 (a) shows a template for a synchronous section. In the template, the value of the expression will be assigned to register Q on the leading edge of signal CLK. We can model the asynchronous sets of resets of registers using the template in Fig. 8.9 (b). There can be more than one test for asynchronous set or reset signals. However, most of the synthesis tools require the test of a signal's edge statement to be the last test of the **if** statement. Fig. 8.9 (b) shows a template of a synchronous section with asynchronous inputs. A group of sequential statements which are not enclosed by an **if** signal'edge detection statement is called a combinational section as shown in Fig. 8.9 (c). All the signals referenced in a combinational section must be listed in the sensitivity list. A RTL process may have multiple synchronous sections and combinational sections.

For example, the following program consists of a sequential section and a combinational section. There are two signal assignment statements within the sequential section which means the synthesizer will generate two flip-flops to store the two signals. The combinational section contains a single signal assignment, "PB_PULSE <= Q1 and (not Q2);". In VHDL, whenever there is a change to signalS CLK, Q1 or Q2, this statement will be evaluated. Fig. 8.10 shows a circuit diagram generated by a synthesizer.

```
entity PULSER is
   port  ( CLK, PB : in  BIT ;
             PB_PULSE : out BIT ) ;
end PULSER ;
architecture BHV of PULSER is
   signal Q1, Q2 :  BIT ;
begin
   process ( CLK, Q1, Q2 )
   begin
      if (CLK'event and CLK = '1') then
         Q1 <= PB ;
         Q2 <= Q1 ;
      end if ;
      PB_PULSE <= (not Q1) nor Q2 ;
   end process ;
end BHV ;
```

```
IF (CLK'event AND CLK = '1') THEN

    Q <= expression ;
    - - CLK is the clock input to reg Q

END IF;
```

(a)

```
IF (async_sig = '1') THEN

    Q <= '0';
    - - active high asynchronous reset

ELSIF (CLK'event AND CLK = '1') THEN

    Q <= expression ;
    - - CLK is the clock input to reg Q

END IF;
```

(b)

```
    - - sequential statements not enclosed by
    - - if CLK'edge detection statement
```

(c)

Figure 8.9 Templates of an RTL process (a) A synchronous section (b) a synchronous section with with asynchronous inputs (c) a combinational section.

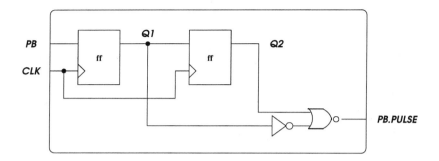

Figure 8.10 A gate level schematic.

8.8 REGISTERS

Various types of registers are used in a circuit. The following VHDL program infers a simple register in Fig. 8.11. The clock pulse enables all flip-flops, so that the information currently available at the four inputs can be transferred into the 4-bit register. The register is preset to an initial value of "1100" asynchronously when signal ASYNC becomes '1',

```
-- 4-bit simple register
signal CLK, ASYNC: BIT ;
signal Din, Dout:   BIT_VECTOR(3 downto 0) ) ;
   . . .
process (CLK, ASYNC)
begin
   if (ASYNC = '1') then
      Dout <= "1100";
   elsif (CLK'event and CLK = '1') then
      Dout <= Din ;
   end if ;
end process ;
```

We may use a *load* signal to control the loading of the input data into a register. When the LOAD input is 1, the inputs are transferred into the register on the next clock pulse. When the LOAD input is 0, the circuit inputs are inhibited and the D flip-flops are reloaded with their present value, thus maintaining the content of the register.

Modeling at the RT Level

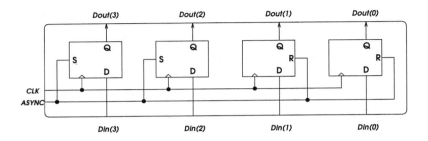

Figure 8.11 A four-bit register.

If there are flip-flops with enable available in the technology library, the synthesis tool may use that to build the circuit. If there are no such a flip-flop, the synthesizer may synthesize a circuit to meet the behavior of the description. using regular flip-flops (Fig. 8.12).

```
-- 4-bit parallel load register
signal CLK, ASYNC, LOAD: BIT ;
signal Din, Dout:  BIT_VECTOR(3 downto 0) ;
 . . .
process (CLK, ASYNC)
begin
   if (ASYNC = '1') then
      Dout <= "1100";
   elsif (CLK'EVENT and CLK = '1') then
      if (LOAD = '1') then
         Dout <= Din ;
      END IF ;
   END IF ;
end process ;
```

A register capable of shifting its binary information either to the right or to the left is called a *shift register*. The simplest possible shift register is one shown in Fig. 8.13. The logical configuration of a shift register consists of a chain of flip-flops connected in cascade, with the output of one flip-flop connected to the input of the next flip-flop. All flip-flops receive a common clock pulse that causes the data to shift from one stage to the next. The following program

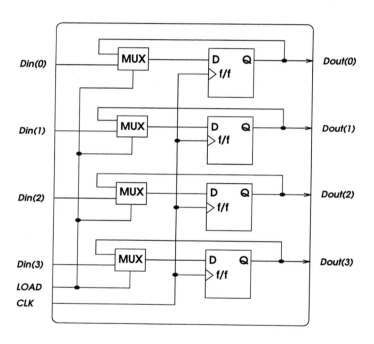

Figure 8.12 A four-bit register with parallel load input.

Modeling at the RT Level

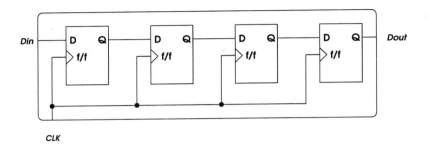

Figure 8.13 A shift register.

segment shows a 4-bit serial-in and serial-out shift register. Noted that REG is declared in the process as a variable. Since it is used before defined, the synthesizer will have to store the value of the variable in flip-flops.

```
-- 4-bit serial-in and serial-out shift register
signal CLK, Din, Dout:   BIT ;
  .  .  .
process ( CLK )
    variable REG      : BIT_VECTOR(3 downto 0);
begin
    if (CLK'event and CLK = '1') then
        REG := Din & REG(3 downto 1);
    end if ;
    Dout <= REG(0) ;
end process ;
```

Shift registers can be used for converting serial data to parallel data, and vice versa. If we have access to all the flip-flop outputs of a shift register, then information entered serially by shifting can be taken out in parallel from the outputs of the flip-flops. Shift registers can also be equipped with input and output terminals for parallel in and out.

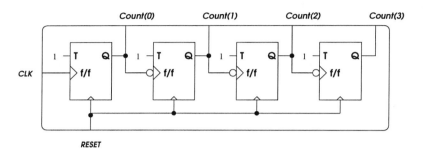

Figure 8.14 A 4-bit binary ripple counter.

8.9 ASYNCHRONOUS COUNTERS

Counters are sequential logic circuits that proceed through a well-defined sequence of states. An asynchronous counter is one whose state changes are no controlled by a synchronizing clock pulse. The most commonly seen asynchronous counter is the binary ripple counters. A binary ripple counter consists of a series connection of complementing flip-flops (T type), with the output of each flip-flop connected to the clock input of the next higher-order flip-flop. The flip-flop holding the least significant bit receives the incoming count pulses. The diagram of a 4-bit binary ripple counter is shown in Fig. 8.14. The small circle of the clock input indicates that the flip-flop complements during a negative-going transition or when the output to which it is connected goes from '1' to '0'.

The following program shows a VHDL program which describes a ffour-bit ripple counter. There are four "if signal's leading edge detection"s statement, each with different clock drivers.

```
-- 4-bit binary counter
signal CLK, RESET: BIT ;
signal COUNT: BIT_VECTOR(3 DOWNTO 0) ;
 . . .
process (CLK, COUNT, RESET)
begin
   if RESET = '1' then
      COUNT <= "0000";
   else
```

```
            if CLK'event and CLK = '1' then
               COUNT(0) <= not COUNT(0) ;
            end if ;
            if COUNT(0)'event and COUNT(0) = '0' then
               COUNT(1) <= not COUNT(1) ;
            end if ;
            if COUNT(1)'event and COUNT(1) = '0' then
               COUNT(2) <= not COUNT(2) ;
            end if ;
            if COUNT(2)'event and COUNT(2) = '0' then
               COUNT(3) <= not COUNT(3) ;
            end if ;
         end if ;
      end process ;
```

8.10 SYNCHRONOUS COUNTERS

If all the flip-flops of a counter are controlled by a common clock signal, it is a synchronous counter. In a synchronous counter, the storage elements simultaneously examine their inputs and determine new outputs. This is the preferred way to build counters.

For example, the following program segment describe a synchronous counter with asynchronous reset, synchronous load, and up or down count. When the *Load* signal equals '1', the count sequence is disabled and the external input is transferred into the counter. If the *Load* signal is '0' and the *Count* signal is '1', the circuit operates as a counter. Depending on the value of signal *Updown*, the clock pulses then cause the state of the register to either count up or count down. If both *Load* and *Count* are '0', the clock pulses do not change the state of the register. Table 8.1 shows the function table of the counter.

```
         -- 4-bit synchronous counter
         signal CLK, RESET : BIT ;
         signal Load, Count, UpDown : BIT ;
         signal DataIn  :  INTEGER range 0 to 15 ;
         signal REG   :  INTEGER range 0 to 15 := 0 ;
            . . .
```

	Input			Mode
RESET	Load	Count	Updown	
H	×	×	×	Asynchronous clear
L	H	×	×	Synchronous load
L	L	L	×	Hold
L	L	H	L	Down count
L	L	H	H	Up count

Table 8.1 Function table of the Synchronous Counter.

```
process (CLK, RESET)
begin
   if RESET = '1' then
      REG <= 0 ;
   elsif CLK'event and CLK = '1' then
      if Load = '1' then
         REG <= DataIn ;
      else
         if Count = '1' then
            if UpDown = '1' then
               REG <= (REG + 1) mod 16 ;
            else
               REG <= (REG - 1) mod 16 ;
            end if ;
         end if ;
      end if ;
   end if ;
end process ;
```

8.11 TRI-STATE BUFFERS

Besides 0 and 1, there is a third signal value in digital systems: the *high-impedance state*, usually denoted by the symbol Z. When a gate's output is in a high-impedance state, it is as though the gate were disconnected from the output. Gates that can be placed in such a state are called *tri-state gates*. They

Modeling at the RT Level

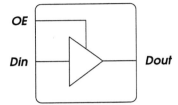

Din	OE	Dout
X	0	Z
0	1	0
1	1	1

Figure 8.15 A tri-state gate.

can produce outputs of the following three values: 0, 1, and Z. In addition to its conventional inputs, a tri-state has one more input called *output enable*. When this input is unasserted, the output is in its high-impedance state and the gate is effectively disconnected from the output wire. When the output enable is asserted, the gate's output is determined by its data inputs.

Among the predefined types of package STANDARD, there is no type to describe the high-impedance value. To specify the high-impedance value in VHDL, multi-value logic type STD_LOGIC in package STD_LOGIC_1164 must be used. When a signal is assigned the value of 'Z', the output of the tri-state gate is disabled. The program segment shows a tri-state gate in Fig. 8.15. There will be a tri-state buffer synthesized betwen signal Din and signal Dout. Signal OE is used to control the state of the tri-state buffer. Note that in the **if** clause, both Din and the condition OE='0' can be full expressions.

```
library IEEE ;
use IEEE.STD_LOGIC_1164.all ;
architecture IMP of Tri_State_Buf is
signal Din, OE, Dout :   STD_LOGIC ;
begin
    . . .
    process(OE, Din)
    begin
      if (OE = '0') then Dout <= 'Z';
      else Dout <= Din;
      end if;
    end process;
end IMP ;
```

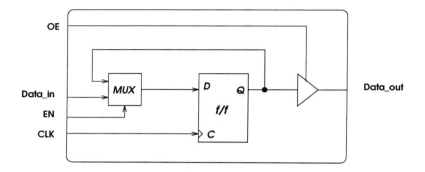

Figure 8.16 A design contains a tri-state gate.

Another handy notation of a tri-state buffer is to use conditional signal assignment as follows.

```
architecture Implement of Tri_State_BUF is
begin
   Dout <= Din when OE = '1' else 'Z' ;
end Implement ;
```

A tri-state gate is combinational logic. If it is described in a synchronous section, the circuit synthesized may not be what a designer expects. The synthesis tool will synthesize a flip-flop to latch the tristate enabled signal before it is connected to the tri-state gate. To prevent such a condition, it is expected that the tri-state gate be specified within a combinational section. For example, the following VHDL program describes a flip-flop with a tristate output shown in Fig. 8.16.

```
library IEEE ;
use IEEE.STD_LOGIC_1164.all ;
entity Tri_DFF is
   port ( CLK,   EN          : in      STD_LOGIC ;
          OE                  : buffer  STD_LOGIC ;
          Data_In             : in      STD_LOGIC ;
          Data_Out            : out     STD_LOGIC ) ;
end Tri_DFF ;
architecture Implement of Tri_DFF is
```

Modeling at the RT Level

```
      begin
        process (CLK, OE)
          variable Temp :  STD_LOGIC ;
        begin
          if (CLK'event and CLK = '1') then
            if EN = '1' then
               Temp := Data_In ;
            end if ;
          end if ;
          if OE = '0' then
            Data_Out <= 'Z' ;
          else
            Data_Out <= Temp ;
          end if ;
        end process ;
      end Implement ;
```

8.12 BUSSES

A bus system can be constructed with three-state gates instead of multiplexers. Fig. 8.17 shows a simple example of a bus system. The outputs of four buffers are connected together to form a single bus line. The control inputs to the buffers (S(0) and S(1) in Fig. 8.17) determine which one of the four normal inputs (R0-R3 in in Fig. 8.17) communicates with the bus line. The designer must guarantee no more than one buffer will be in the active state at any given time. The connected buffers must be controlled so that only one three-state buffer has access to the bus line while all other buffers are maintained in a high-impedance state. The following program describes such a simple bus system.

```
      library IEEE ;
      use IEEE.STD_LOGIC_1164.all ;
      entity BUS_EX is
         port ( S : in  STD_LOGIC_VECTOR(1 downto 0) ;
                OE : buffer STD_LOGIC_VECTOR(3 downto 0) ;
                R0, R1, R2, R3 :  in STD_LOGIC_VECTOR(7 downto 0) ;
```

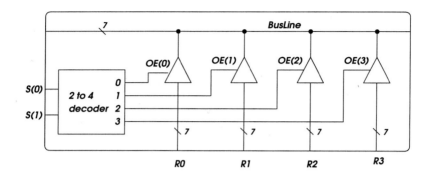

Figure 8.17 A bus system.

```
            BusLine :   out STD_LOGIC_VECTOR(7 downto 0) ) ;
END BUS_EX ;
architecture Implement of BUS_EX is
begin
   process (S)
   begin
      case (S) is
         when "00" => OE <= "0001" ;
         when "01" => OE <= "0010" ;
         when "10" => OE <= "0100" ;
         when "11" => OE <= "1000" ;
         when others => null ;
      end case ;
   end process ;
   BusLine <= R0 when OE(0) = '1' else 'ZZZZZZZ' ;
   BusLine <= R1 when OE(1) = '1' else 'ZZZZZZZ' ;
   BusLine <= R2 when OE(2) = '1' else 'ZZZZZZZ' ;
   BusLine <= R3 when OE(3) = '1' else 'ZZZZZZZ' ;
end Implement ;
```

Normally, simultaneous assignment to a signal such as BusLine in the previous example is not allowed at the architectural level unless a resolution function is provided. However, in VHDL data types STD_LOGIC or STD_LOGIC_VECTOR

Modeling at the RT Level

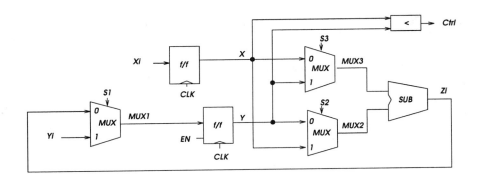

Figure 8.18 A data path.

are already resolved (a resolution function is provided in STD_LOGIC_1164). Therefore, signals declared with these two data types can have multiple drivers.

8.13 NETLIST OF RTL COMPONENTS

A data path usually consists of a netlist of RTL components such as function units, multiplexers, comparators, registers, *etc.* Fig. 8.18 shows an example data path. We can describe each component as a single process and connect them using signals. Alternatively, we can describe all of them in one process.

To describe all the components in one process, we need to know how VHDL simulator works. Since the statements within a process are executed sequentially, the inputs of a component should have been evaluated by all its predecessor components. Or otherwise, the simulator will get the old value of its predecessor component. Therefore, if we start with the description of flip-flops, then we can describe the two multiplexors and the comparator because they are connected to the outputs of the registers. Then, we can describe the subtractor, and finally the multiplexor. The following program segment describe the circuit in Fig. 8.18.

```
-- a netlist of RTL components
signal CLK, EN, S1, S2, S3 :  BIT ;
signal Xi, Yi, Zi :   INTEGER range 0 to 255
signal Ctrl:  BOOLEAN ;
```

```
     . . .
    begin
       process (CLK, Xi , Yi , S1 , S2, S3)
          variable X, Mux0, Mux1, Mux2 :   INTEGER range 0 to 255 ;
       begin
          if (CLK'event and CLK = '1') then    -- registers
             X := Xi ;
             if EN then Y := Zi ;   end if ;
          end if ;
          Ctrl <= (X < Y) ;                    -- comparator
          case S2 is                           -- multiplexor
             when '0' => Mux2 := Y ;
             when '1' => Mux2 := X ;
          end case ;
          case S3 is                           -- multiplexor
                when '0' => Mux3 := X ;
                when '1' => Mux3 := Y ;
          end case ;
          Zi <= Mux3 - Mux2;                   -- subtractor
          case S1 is                           -- multiplexor
             when '0' => Mux1 := Zi ;
             when '1' => Mux1 := Yi ;
          end case ;
       end process ;
```

8.14 SUMMARY

1. In this chapter, we discuss the modeling of an RTL circuit. A Register Transfer Level (RTL) design is characterized by a set of registers connected by combinational logic blocks. The behavior of an RTL circuit can be modeled by the behavior constructs of VHDL.

Modeling at the RT Level 157

2. A process which does not contain **if** signal leading edge (or falling edge) statements or **wait** signal'event statements is called a *combinational process*.

3. A combinational process is usually synthesized as a combinational circuit. Latches will be generated when there are incompletely specified conditional expressions in a combinational description.

4. A process with **if** signal leading edge (or falling edge) statements or **wait** signal'event statements is called a *clocked process*. An edge triggered flip-flop will be generated from a VHDL description if a signal assignment (or possibly variable assignment) is executed on the leading (or on the falling) edge of another signal.

5. To specify the high-impedance value in VHDL, multi-value logic type STD_LOGIC in package STD_LOGIC_1164 must be used. When a signal is assigned the value of 'Z', the output of the tri-state gate is disabled.

Exercises

1. Write a VHDL description for an 8-bit non-priority encoder.

2. Write a VHDL description for an 8-bit priority encoder.

3. Write a VHDL program to convert a BCD number (4-bit) to a 7-segment display.

4. A T-type (toggle) flip-flop has a single data input denoted T and is characterized by the fact that $T = 1$ causes the flip-flop to change state, while $T = 0$ retains the current state. Describe the behavior of a T-type flip-flop in VHDL.

5. Write a VHDL program to perform an addition of two BCD numbers.

6. A 4-bit shift register has a *load*, *mode*, a *serial_input*, and *clock* inputs, four lines of *parallel_in* and four lines of *parallel_out*. When *load* is high, on the falling edge of the *clock* the *parallel_in* is loaded into the shift register. When *mode* is high, the shift register is in the right-shift mode and on the falling edge of the *clock*, the *serial_input* is clocked into the most significant bit. When *mode* is low, the shift register is in the left-shift mode and on the falling edge of the *clock*, the *serial_input* is clocked into the least significant bit. Write an RTL design description for the shifter.

7. Consider the following VHDL program. Draw a RT level schematic for the design.

```
SIGNAL OU: BIT;
SIGNAL CLK, SET: BIT;
BEGIN
    PROCESS(CLK, SET)
    VARIABLE A: BIT;
    BEGIN
        IF(SET = '1') THEN
            OU <= '1';
        ELSE
            IF(CLK'event and CLK='1') THEN
                A := not A;
            END IF;
            IF(A'event and A='1') THEN
```

Modeling at the RT Level

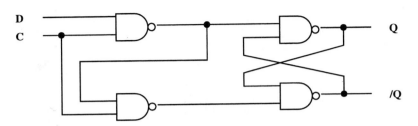

Figure X8.1

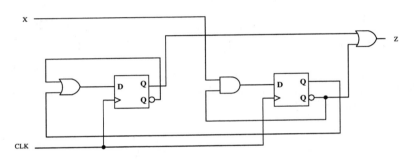

Figure X8.2

```
            OU <= not OU;
        END IF;
      END IF;
   END PROCESS;
END;
```

8. Describe the D latch in Figure X8.1 in VHDL.

9. Describe the clock synchronous state machine in Figure X8.2 in VHDL at the RT level.

10. Describe the clock synchronous state machine in Figure X8.3 in VHDL at the RT level.

11. Shown in Figure X8.4 is the circuit schematic of a simple computer. Describe the data path in VHDL.

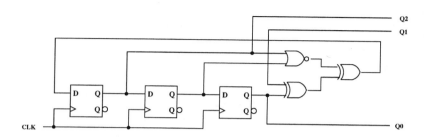

Figure X8.3

12. Consider the following VHDL program. Suppose you have a D flip-flop which has asynchronous preset and reset inputs. What is the best hardware implementation of the VHDL description? Now if you don't have such a flop-flop, but you have a flip-flop which has only one asynchronous input which can be programmed to be preset or reset. What is the best hardware implementation of the VHDL description?

```
entity RLatch is
   port ( S,X       : in  BIT ;
          Async     : in  BIT ;
          Din       : in  BIT ;
          Dout      : out BIT ) ;
end RLatch;
architecture Implement of Latch is
begin
   process (S, Async, X, Din)
   begin
      if (ASYNC = '1') then
        Dout <= X;
      elsif (if CLK'event and CLK = '1') then
        Dout <= Din;
      end if;
   end process ;
end Implement ;
```

Modeling at the RT Level

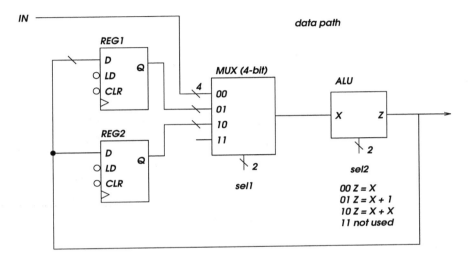

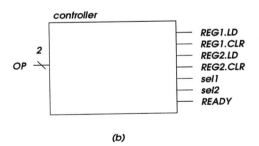

Figure X8.4

9

MODELING AT THE FSMD LEVEL

A digital design is conceptually divided into two parts – a controller and a datapath. The relationship between the control logic and the datapath in a digital system is shown in Fig. 9.1. The datapath manipulates data in registers (or memories) according to the commands from the controller. The controller provides the datapath with the appropriate commands at every moment in time so that the datapath properly implements the specified functions and produces the required external output signals. The controller uses status conditions from the datapath to serve as decision variables for determining the sequence of state transitions.

The controller that generates the control commands to the datapath is a sequential circuit whose internal states determine the control commands of the system. From the present state (PS), depending on status information and external inputs, the sequencer goes to the next state (NS) to initiate other operations and generates necessary outputs. The basic model of a sequencer consists of a combinational circuit and a state register. The state register keeps the current state, and the combinational circuit produces the output commands and next state (NS) based on the present state (PS) and external inputs (including status information). A sequential circuit which is implemented in a fixed number of possible states is called a finite state machine (FSM). Finite state machines are critical for realizing the control and decision-making logic in digital systems.

In this chapter, we will start with a VHDL design description for two basic models of FSMs that are well-known: namely the Moore machine, and the Mealy machine. Then, we will use VHDL to describe a circuit which contains a data path and a control unit. A flow chart is a convenient way to specify the

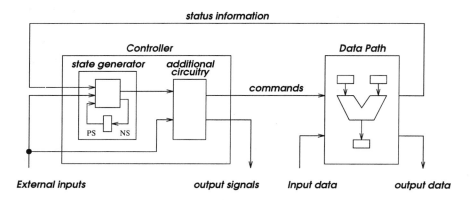

Figure 9.1 Interaction of a Controller and a datapath.

sequence of procedure steps and decision paths for an algorithm. A flow chart for a hardware algorithm translates word statements to an information diagram that lists the sequence of operations together with the conditions necessary for their execution. A special flow chart that has been developed specifically to define finite state machines is called an *algorithmic state machine* (ASM) chart. FSMD (FSM with a datapath) is an extension of an FSM with the capability of describing integer variable operations within a state description. ASM chart can be easily applied to describe FSMD.

9.1 MOORE MACHINES

In the Moore model of sequential circuits, the outputs are functions of the present state only. Fig. 9.2 shows the block diagram of a Moore machine. A combinational logic block maps the inputs and the current state into the inputs of the flip-flops which store the next state. The outputs are computed by a combinational logic block whose only inputs are the flip-flops' state outputs (see Fig. 9.2). In a Moore model, the outputs of the sequential circuit are synchronized with the clock because they depend only on flip-flop outputs that are synchronized with the clock.

A Moore model of a sequential circuit can be represented by a state diagram with the circles marked with the current state and output, and directed edges are marked with input only. Fig. 9.3 shows an example of a state transition

Modeling at the FSMD Level

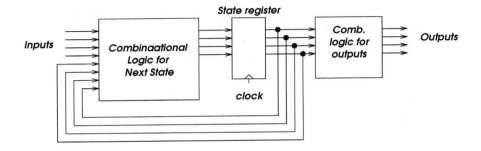

Figure 9.2 A block diagram of a Moore machine.

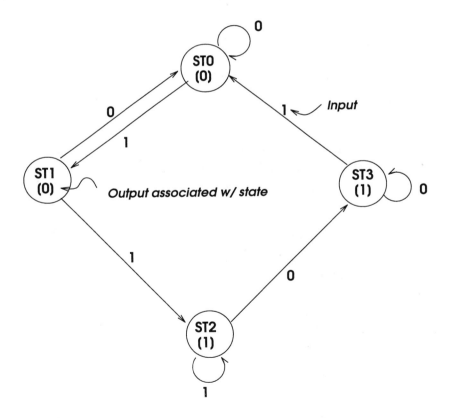

Figure 9.3 A state transition diagram of a Moore machine.

diagram for a Moore machine. The example below shows a VHDL description for the Moore machine. Similar to an RTL description, we can divide a process of a state machine into a sequential section and a combinational section. The asynchronous reset signal initializes some registers and resets the finite state machine to its initial state asynchronously. At each leading edge of signal CLK, the value of the next state is assigned into the current state. The combinational section describes the combinational logic for generating the next state and the outputs of the design. Since the outputs are computed based on the current state only, the assignments of output signals cannot be dependent on the input signals.

```
entity MOORE is
   port ( CLK      : in  BIT ;
          RST      : in  BIT ;
          I        : in  BIT ;
          O        : out BIT ) ;
end MOORE ;
architecture Implement of MOORE is
begin
   process (CLK, RST, I)
      type STATE_TYPE is (ST0,ST1,ST2,ST3);
      variable State,Next_State :   STATE_TYPE ;
      -- mebs state_var State,Next_State
   begin
      if (RST = '1') then
         O <= '0' ;
         State := ST0 ;
      elsif (CLK'event and CLK = '1') then
         State := Next_State ;
      end if ;
      case State is
         when ST0 =>
            O <= '0' ; -- output only depends on present state
            if (I = '0') then   Next_State := ST0 ;
            else                Next_State := ST1 ;
            end if ;
         when ST1 =>
            O <= '0' ;
```

```vhdl
                    if (I = '0') then  Next_State := ST0 ;
                    else               Next_State := ST2 ;
                    end if ;
                when ST2 =>
                    O <= '1' ;
                    if (I = '0') then  Next_State := ST3 ;
                    else               Next_State := ST2 ;
                    end if ;
                when ST3 =>
                    O <= '1' ;
                    if (I = '0') then  Next_State := ST3 ;
                    else               Next_State := ST0 ;
                    end if ;
            end case ;
        end process ;
    end Implement ;
```

9.2 ASYNCHRONOUS MEALY MACHINES

A block diagram of a Mealy model is shown in Fig. 9.4. In the Mealy model, the outputs are functions of both the present state and current inputs. The outputs may change if the inputs change during the clock pulse period. Because of this, the outputs may have momentary false values because of the delay encountered from the time that the inputs change and the time that the flip-flop outputs change. In order to synchronize an asynchronous Mealy machine, the inputs of the sequential circuit must be synchronized with the clock and the outputs must be sampled only during the clock-pulse transition.

In a state transition diagram representation of a Mealy machine, both the input and output values are included along the directed lines between the circles. Fig. 9.5 shows a sample state diagram for a Mealy machine. The following VHDL program describes the asynchronous Mealy machine in Fig. 9.5. A process describing a Mealy machine can be divided into a synchronous section and a combination section. The asynchronous reset in the synchronous section initializes some registers and resets the finite state machine to its initial state. At each leading edge of the signal CLK, the value of the next state is assigned

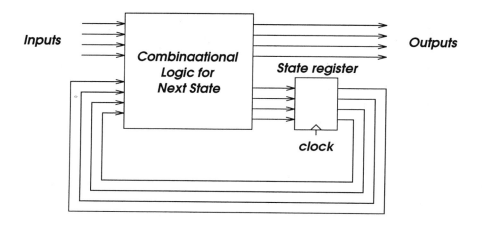

Figure 9.4 A block diagram of a Mealy machine.

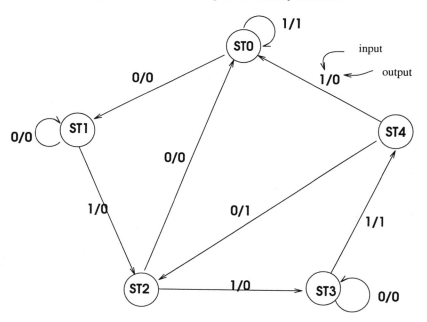

Figure 9.5 A state transition diagram of a Mealy machine

Modeling at the FSMD Level 169

into the current state. The combinational section describes the combinational logic for generating the next state and the outputs of the design. The output signals are generated based on the value of the input signals.

```
entity ASYNC_MEALY is
    port ( CLK            : in  BIT ;
           RST            : in  BIT ;
           I              : in  BIT ;
           O              : out BIT ) ;
end ASYNC_MEALY ;
architecture Implement of ASYNC_MEALY is
begin
    process (CLK, RST, I)
        type STATE_TYPE is (ST0,ST1,ST2,ST3,ST4) ;
        variable State,Next_State :   STATE_TYPE ;
        -- mebs state_var State,Next_State
    begin
        if (RST = '1') then
            State := ST0;
        elsif (CLK'event and CLK = '1') then
            State := Next_State ;
        end if ;
        case STATE is
            when ST0 =>
                if (I = '0') then  O <= '0';  Next_State := ST1 ;
                else               O <= '1';  Next_State := ST0 ;
                end if ;
            when ST1 =>
                if (I = '0') then  O <= '0';  Next_State := ST1 ;
                else               O <= '0';  Next_State := ST2 ;
                end if ;
            when ST2 =>
                if (I = '0') then  O <= '0';  Next_State := ST0 ;
                else               O <= '0';  Next_State := ST3 ;
                end if ;
            when ST3 =>
```

```
                    if (I = '0') then    O <= '0';   Next_State := ST3 ;
                    else                 O <= '1';   Next_State := ST4 ;
                    end if ;
                 when ST4 =>
                    if (I = '0') then    O <= '1';   Next_State := ST2 ;
                    else                 O <= '0';   Next_State := ST0 ;
                    end if ;
              end case ;
           end process ;
        end Implement ;
```

The previous VHDL description can be rewritten by using some default assignments before the case statement of the combinational section. The default assignments assign signal '0' to O, and ST0 to Next_State. If this signal has not been re-assigned in the case statement, it will stay the same value. However, if it is re-assigned, the later assignment will over-ride the default assignments.

```
        architecture Implement1 of ASYNC_MEALY is
        begin
           process (CLK, RST, I)
              type STATE_TYPE is (ST0,ST1,ST2,ST3,ST4) ;
              variable State,Next_State :   STATE_TYPE ;
              -- mebs state_var State,Next_State
           begin
              if (RST = '1') then
                 State := ST0 ;
              elsif (CLK'event and CLK = '1') then
                 State := Next_State ;
              end if ;
              -- default statements
              O <= '0' ;
              Next_State := ST0 ;
              case State is
                 when ST0 =>
                    if (I = '0') then               Next_State := ST1 ;
                    else             O <= '1';
                    -- no state assignment here, using default
```

Modeling at the FSMD Level 171

```
                end if ;
            when ST1 =>
                if (I = '0') then           Next_State := ST1 ;
                else                        Next_State := ST2 ;
                end if ;
            when ST2 =>
                if (I = '1') then           Next_State := ST3 ;
                end if ;
            when ST3 =>
                if (I = '0') then           Next_State := ST3 ;
                else            O <= '1';   Next_State := ST4 ;
                end if ;
            when ST4 =>
                if (I = '0') then O <= '1'; Next_State := ST2 ;
                end if ;
        end case ;
    end process ;
end Implement1 ;
```

9.3 SYNCHRONOUS MEALY MACHINES

In an asynchronous Mealy model, the outputs respond asynchronously with the inputs. This may produce glitches in the circuit outputs. An alternative synchronous design style for Mealy machines is to break the direct connection between inputs and outputs by introducing storage elements. One way to do this is to synchronize the outputs with output flip-flops. Fig. 9.6 shows a block diagram of synchronous Mealy machine. The flip-flops are clocked with the same edge as the state register. Since all the signal outputs and next state outputs are synchronized with the clock, we can describe the state transition function in the synchronous section. The example below shows a VHDL program which describes a synchronous Mealy state machine of the state transition diagram in Fig. 9.5. Note that there is no combinational section in the program. The primary outputs and the next state outputs are registered. Therefore, they are specified in the synchronous section.

```
    entity SYNC_MEALY is
        port ( CLK               : in  BIT ;
```

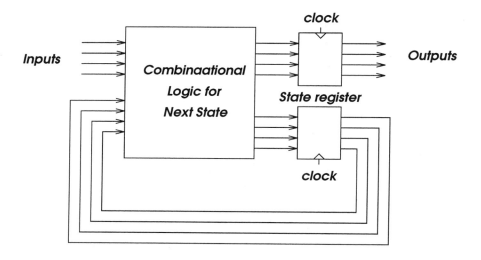

Figure 9.6 A block diagram of a Synchronous Mealy machine.

```
           RST            : in  BIT ;
           I              : in  BIT ;
           O              : out BIT ) ;
end SYNC_MEALY ;
architecture Implement of SYNC_MEALY is
begin
   process (CLK, RST, I)
      type State_Type is (ST0,ST1,ST2,ST3,ST4) ;
      variable State    :  State_Type ;
      -- mebs state_var State
   begin
      if (RST = '1') then
         State := ST0 ;
      elsif (CLK'event and CLK = '1') then
         case State is
            when ST0 =>
               if (I = '0') then  O <= '0';  State := ST1 ;
               else               O <= '1';  State := ST0 ;
```

Modeling at the FSMD Level

```
              end if ;
           when ST1 =>
              if (I = '0') then   O <= '0';   State := ST1 ;
              else                O <= '0';   State := ST2 ;
              end if ;
           when ST2 =>
              if (I = '0') then   O <= '0';   State := ST0 ;
              else                O <= '0';   State := ST3 ;
              end if ;
           when ST3 =>
              if (I = '0') then   O <= '0';   State := ST3 ;
              else                O <= '1';   State := ST4 ;
              end if ;
           when ST4 =>
              if (I = '0') then   O <= '1';   State := ST2 ;
              else                O <= '0';   State := ST0 ;
              end if ;
         end case ;
      end if ;
   end process ;
end Implement ;
```

9.4 SEPARATION OF FSM AND DATAPATH

A digital design is conceptually divided into two parts – a controller and a datapath as shown in Fig. 9.1. The controller is an FSM which issues commands to the datapath based on the current state and the external inputs. The datapath consists of a netlist of function units, multiplexers and registers. The VHDL description of a datapath in the RT level has been discussed in chapter 8.

In the following, we show a digital design which is used to solve the "greatest common divisor of two eight-bit numbers". The design consists of a datapath and a controller. The description of the controller is based on the Mealy model of FSM. Fig. 9.7 shows the block diagram of the GCD calculator. The following example shows a VHDL description which describes the data path and the

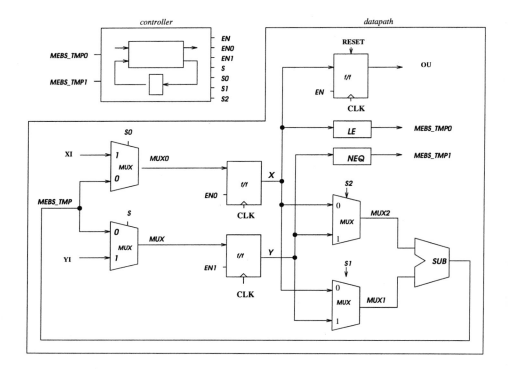

Figure 9.7 A block diagram of the GCD calculator.

controller in two separate processes. Process GCD_DP describes the datapath and process GCD_FSM describes the controller. On top of the two processes, we define a block in which we declare signals which are either commands to the data path or status information for the controller.

```
entity GCD is
   port ( CLK, RESET   : in  BIT ;
          START        : in  BIT ;
          XI, YI       : in  INTEGER range 0 to 255 ;
          OU           : out INTEGER range 0 to 255 ) ;
end GCD ;
architecture RTL of GCD is
begin
```

```
Main : block
  type   Sel_Type_2 is (SEL0 ,SEL1) ;
  signal MEBS_TMP0, MEBS_TMP1     : BOOLEAN ;
  signal EN, EN0, EN1             : BOOLEAN ;
  signal S, S0, S1, S2            : Sel_Type_2 ;
begin
  GCD_DP :
  process (CLK, RESET, XI , YI , S  , S0 , S1 , S2)
      variable X, Y, MEBS_TMP : INTEGER range 0 to 255;
      variable MUX, MUX0, MUX1, MUX2: INTEGER range 0 to 255;
  begin
     if (RESET = '1') then
        OU <= 0 ;
     elsif (CLK'event and CLK = '1') then
        if EN then      OU <= X ;       end if ;
        if EN0 then     X := MUX0 ;     end if ;
        if EN1 then     Y := MUX ;      end if ;
     end if;
     MEBS_TMP0 <= (X < Y) ;
     MEBS_TMP1 <= (X /= Y) ;
     case S1 is                 -- multiplexor
        when SEL0 => MUX1 := Y ;
        when SEL1 => MUX1 := X ;
     end case ;
     case S2 is                 -- multiplexor
        when SEL0 => MUX2 := X ;
        when SEL1 => MUX2 := Y ;
     end case ;
     MEBS_TMP := (MUX1 - MUX2) mod 256 ;
     case S0 is                 -- multiplexor
        when SEL0 => MUX0 := MEBS_TMP ;
        when SEL1 => MUX0 := XI ;
     end case ;
     case S is                  -- multiplexor
        when SEL0 => MUX := MEBS_TMP ;
```

```
         when SEL1 => MUX := YI ;
      end case ;
   end process GCD_DP ;
   GCD_FSM :
   process (CLK, RESET, START, MEBS_TMP0, MEBS_TMP1)
      type S_Type is (ST0 , ST1) ;
      variable State,Next_State :   S_Type ;
      -- mebs state_var State, Next_state
   begin
      if (RESET = '1') then
         State := ST0 ;
      elsif (CLK'event and CLK = '1') then
         State := Next_State ;
      end if ;
      S  <= SEL0 ;
      S0 <= SEL0 ;
      S1 <= SEL0 ;
      S2 <= SEL0 ;
      -- mebs dont_care S, S0, S1, S2
      EN  <= FALSE ;
      EN0 <= FALSE ;
      EN1 <= FALSE ;
      case State is
         when ST0 =>
            if (START = '1') then
               EN0 <= TRUE ;
               EN1 <= TRUE ;
               S1  <= SEL1 ;
               S2  <= SEL1 ;
               Next_State := ST1 ;
            else
               Next_State := ST0 ;
            end if ;
         when ST1 =>
            if (MEBS_TMP1 and (not MEBS_TMP0)) then
```

```
                    EN0 <= TRUE ;
                    S   <= SEL1 ;
                    S0  <= SEL1 ;
                    S2  <= SEL0 ;
                    Next_State := ST1 ;
                elsif (MEBS_TMP1 and MEBS_TMP0) then
                    EN1 <= TRUE ;
                    S   <= SEL0 ;
                    S0  <= SEL0 ;
                    S1  <= SEL0 ;
                    Next_State := ST1 ;
                else
                    EN  <= TRUE ;
                    Next_State := ST0 ;
                end if ;
            end case ;
        end process GCD_FSM ;
    end block Main ;
end RTL ;
```

9.5 AN FSM WITH A DATAPATH (FSMD)

So far the capability of an FSM mentioned above has been very restricted. It cannot represent variables and their operations. It cannot represent storage elements (registers) except the state registers. It works well for a design with a few to several hundred states. Beyond several hundred states, the model becomes incomprehensible to human designers. In case we have to represent an 8-bit integer variable, we may need 2^8 states to represent different combinations of the variable.

An **FSM with a datapath** (FSMD) is an extension of a traditional FSM in which storage variables can be declared. Within a state expression, we can perform comparison, arithmetic or logic operations on these variables. Each variable may replace thousands of different states in a traditional FSM. With the introduction of variables and variable computation in a state description, we have more flexibility in modeling a digital circuit.

Present state	Input	Next state	Output
$Q_1 Q_0$	$Count$	$Q_1 Q_0$	y
$s_0 = 00$	1	$s_1 = 01$	0
$s_1 = 01$	1	$s_2 = 10$	0
$s_2 = 10$	1	$s_0 = 00$	1
don't care	0	$s_0 = 00$	0

(a)

Present state	Input	Next state	Output
s_0	$(Count = 1)$ and $(x \neq 2)$	s_0	$x = x+1, y = 0$
	$(Count = 1)$ and $(x = 2)$		$x = 0, y = 1$
	$(Count = 0)$		$x = 0, y = 0$

(b)

Table 9.1 Comparison between FSM and FSMD (a) FSM (b) FSMD

Take the modulo-3 divider as an example. Table 9.1(a) shows a state transition table of an FSM description, and Table 9.1(b) shows a state transition table of an FSMD description. The FSMD design contains a 2-bit storage variable x. x can be compared ($x < 2$ or $x = 2$) and computed ($x+1$) in a state description. On the other hand, the input of a traditional FSM must be primary inputs and the output must be primary outputs.

The following program shows the difference between the two notations for the modulo-3 divider.

```
entity M3 is
  port (RESET,CLOCK : in  BIT ;
        COUNT       : in  BIT ;
        Y           : out BIT ) ;
end M3 ;
architecture FSM of M3 is
begin
  process (RESET,CLOCK)
    type State_Type is (ST0, ST1, ST2) ;
    variable STATE,NEXT_STATE : State_Type ;
```

```vhdl
    begin
      if (RESET = '1') then
        STATE := ST0 ;
        NEXT_STATE := ST0 ;
      elsif (CLOCK'event and CLOCK = '1') then
        STATE := NEXT_STATE ;
      end if ;
      case STATE is
        when ST0 =>
          if (COUNT = '1') then
            Y <= '0'; NEXT_STATE := ST1 ;
          else
            Y <= '0'; NEXT_STATE := ST0 ;
          end if ;
        when ST1 =>
          if (COUNT = '1') then
            Y <= '0'; NEXT_STATE := ST2 ;
          else
            Y <= '0'; NEXT_STATE := ST0 ;
          end if ;
        when ST2 =>
          if (COUNT = '1') then
            Y <= '1'; NEXT_STATE := ST0 ;
          else
            Y <= '0'; NEXT_STATE := ST0 ;
          end if ;
      end case ;
    end process ;
end FSM ;
architecture FSMD of M3 is
begin
  process (RESET,CLOCK)
    variable X : INTEGER range 0 to 2 ;
  begin
    if (RESET = '1') then
```

```
              X := 0 ;
        elsif (CLOCK'event and CLOCK = '1') then
          if (COUNT = '1') then
            if (X /= 2) then
              Y <= '0'; X := X + 1 ;
            else
              Y <= '1'; X := 0 ;
            end if ;
          else
            Y <= '0'; X := 0 ;
          end if ;
        end if ;
      end process ;
    end FSMD ;
```

With the capability of describing variables and variable computations in a state description, traditional state transition diagrams and state transition tables become insufficient to represent a FSMD. A more powerful representation is required. The behavior of a FSMD can be represented as a flow-chart-like description such as algorithmic state machine (ASM) notation. An ASM chart is represented by three basic elements: the state box, the decision box, and the conditional output box as shown in Fig. 9.8. A state in the control sequence is indicated by a state box (a rectangular box). The decision box describes the effect of an input on the system. It has a diamond-shaped box with two or more exit paths. The conditional output box is represented as an oval shaped box. The input path of a conditional box must come from one of the exit paths of a decision box. The operations listed inside the conditional box will be executed during a given state provided that the input condition is satisfied.

Fig. 9.9 shows the GCD calculator in an FSMD representation. The flow chart represents a state by a state behavior of calculating the GCD of the two numbers. In the representation, the input variables, XI and YI are two 8-bit numbers. Also, subtraction operations are specified within the state description. Hence this is a FSMD description.

The VHDL description of an FSMD level design can be structured in a very similar way as that of a synchronous Mealy machine. It contains a synchronous section (all the outputs will be latched by flip-flops). The test of an asynchronous reset determines the actions that will be triggered by the asynchronous reset

Modeling at the FSMD Level

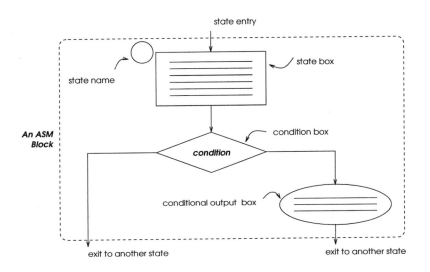

Figure 9.8 Elements of the ASM notation.

signal. The synchronous section describes the computation within the state and the state transition governed by the clock pulse. We are allowed to declare integers and we can perform numeric computation inside a state description. The following example shows a VHDL program for the GCD-calculator at the FSMD level.

```
architecture FSMD of GCD is
begin
   Main :
   process (RESET, CLOCK)
      variable X, Y: INTEGER range 0 to 255 ;
      type S_Type IS (ST0 , ST1 );
      variable State         :  S_Type := ST0 ;
      variable TMP_0, TMP_1  :  BOOLEAN ;
      -- mebs state_var State
   begin
      if (RESET = '1') then
         OU <= 0 ;
         State := ST0 ;
```

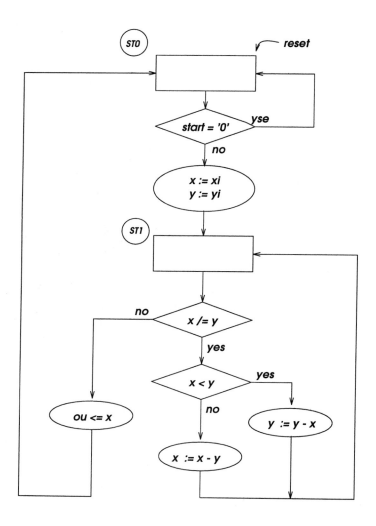

Figure 9.9 FSMD of the GCD calculator.

```
            elsif (CLOCK'event and CLOCK = '1') then
                case State is
                    when ST0 =>
                        if (START = '1') then
                            X := XI ;
                            Y := YI ;
                            State := ST1 ;
                        else
                            State := ST0 ;
                        end if ;
                    when ST1 =>
                        if (X /= Y) then
                            if (X < Y) then
                                Y := Y - X ;
                            else
                                X := X - Y ;
                            end if ;
                            State := ST1 ;
                        else
                            OU <= X ;
                            State := ST0 ;
                        end if ;
                end case ;
            end if ;
    end process Main ;
end FSMD ;
```

9.6 COMMUNICATING FSMS

Fig. 9.10(a) shows ASMs for two communicating finite state machines (FSMs). This is an example of a two-way handshaking protocol which can be seen in many communication applications. Assume both are positive edge-triggered synchronous systems and the output from each state machine is the input to the other. The interaction between these two machines is illustrated by the timing diagram in Fig. 9.10(b). Note that both FSMs are using the same clock signal.

The following program describes the two communicating FSMs in Fig. 9.10(a). The initial state of process FSM_A is state ST0. When it sees the *ack* signal becomes 0, it sends a READY signal to FSM_B and transfers to ST1. At state ST1, it waits until it sees the *ack* signal changes to 1. Then it resets the READY signal and transfers back to ST0.

```
architecture Example of HandShake is
    signal READY, ACK    : BIT := '0' ;
begin
    FSM_A :
    process (RESET, CLK)
        type S_Type    is (ST0, ST1) ;
        variable State :  S_Type ;
        -- mebs state_var State
    begin
        if (RESET = '1') then
            READY <= '0' ;
            STATE := ST0 ;
        elsif (CLK'EVENT and CLK = '1') then
            case STATE is
                when ST0 =>
                    if (ACK = '0') then
                        READY <= '1' ;
                        -- send data to communication wire
                        STATE := ST1 ;
                    else
                        STATE := ST0 ;
                    end if ;
                when ST1 =>
                    if (ACK = '1') then
                        READY <= '0' ;
                        STATE := ST0 ;
                    else
                        STATE := ST1 ;
                    end if ;
            end case ;
        end if ;
```

Modeling at the FSMD Level 185

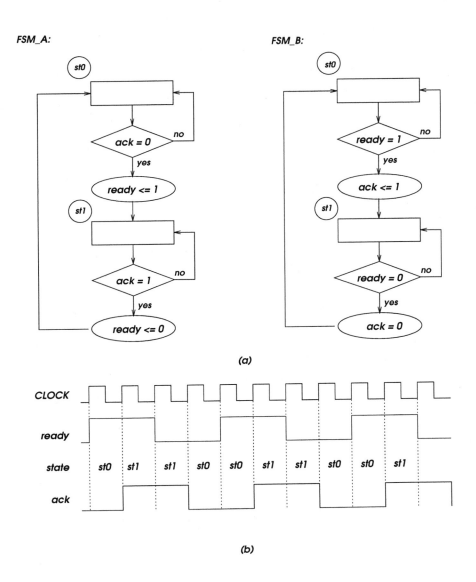

Figure 9.10 Communicating FSMs.

```
        end process FSM_A ;
     FSM_B :
     process (RESET, CLK)
        type S_Type    is (ST0, ST1) ;
        variable STATE : S_Type ;
        -- mebs state_var state
     begin
        if (RESET = '1') then
           ACK <= '0' ;
           STATE := ST0 ;
        elsif (CLK'EVENT and CLK = '1') then
           case STATE is
              when ST0 =>
                 if (READY = '1') then
                    -- read data from communication wire
                    ACK <= '1' ;
                    STATE := ST1 ;
                 else
                    STATE := ST0 ;
                 end if ;
              when ST1 =>
                 if (READY = '0') then
                    ACK <= '0' ;
                    STATE := ST0 ;
                 else
                    STATE := ST1 ;
                 end if ;
           end case ;
        end if ;
     end process FSM_B ;
  end Example ;
```

9.7 SUMMARY

1. A digital design is conceptually divided into two parts – a controller (FSM) and a datapath. The datapath manipulates data in registers (or memories) according to the commands from the controller.

2. In a Moore model FSM, the outputs are functions of the present state only.

3. In an asynchronous Mealy model FSM, the outputs are functions of both the present state and current inputs. The outputs may change if the inputs change during the clock pulse period.

4. In an asynchronous Mealy FSM, the outputs respond asynchronously with the inputs. In other words, the outputs may change if the inputs change during the clock pulse period.

5. A synchronous Mealy FSM synchronizes the outputs by inserting flip-flops for the output signals.

6. FSMD is an extension of the traditional FSM. It allows the use of status variables and storage elements in a state description. Each variable in an FSMD can replace thousands of different states in an FSM.

Exercises

1. Write a VHDL program for the following Moore machine. A finite state machine has one input and one output. The output becomes 1 and remains 1 thereafter when at least two 0's and at least tow 1's have occurred as inputs, regardless of the order of occurrence. Show the simulation waveform.

2. Write a VHDL program for the following asynchronous Mealy machine. A two-input Mealy machine that produces a 1 at its single output when the values of the two inputs differ at the time of the previous clock pulse. Show the simulation waveform.

3. A sequential circuit has two inputs and two outputs. The inputs (X_1, X_2) represent a 2-bit binary number, N. If the present value of N is greater than the previous value, then $Z_1 = 1$. If the present value of N is less than the previous value, then $Z_2 = 1$. Otherwise, Z_1 and Z_2 are 0. Write a VHDL program for the machine.

4. Build a state machine which emits the following one of two BCD number sequences, depending on the value of a control variable MODE. Each sequence has a cycle of four digits:

   When MODE = '0', the sequence is 0, 1, 2, 3, 0, 1, ...
   When MODE = '1', the sequence is 9, 8, 7, 6, 9, 8, ...

   Show the simulation waveform.

5. A sequential circuit has two inputs and two outputs. The inputs (X_1, X_2) represent a 2-bit binary number, N. If the present value of N is greater than the previous value, then $Z_1 = 1$. If the present value of N is less than the previous value, then $Z_2 = 1$. Otherwise, Z_1 and Z_2 are 0. Write a VHDL program for the machine.

6. Write a VHDL program for a state machine that samples a continuous stream of synchronized data on an input line X. The state machine is to an output Z any time the sequence ...0110... occurs. Consider that the sequence may be overlapping. For example,

$$X = 0011001101100110$$
$$Z = 0000100010010001$$

7. Develop a design description for a four-bit up/down, modulo 10 counter with the following function table.

s_1	s_0	Mode
0	0	Up
0	1	Down
1	0	Modulo 10
1	1	Modulo 10

8. The problem is to design a state machine to control the tail lights of a car. There are three lights on each side – LC LB LA || RA RB RC. For turns they operate in sequence to show the turning direction as illustrated in the following:

```
    left turn                right turn
    ---------                ----------

          LA                 RA
       LB LA                 RA RB
    LC LB LA                 RA RB RC
```

Modeling at the FSMD Level

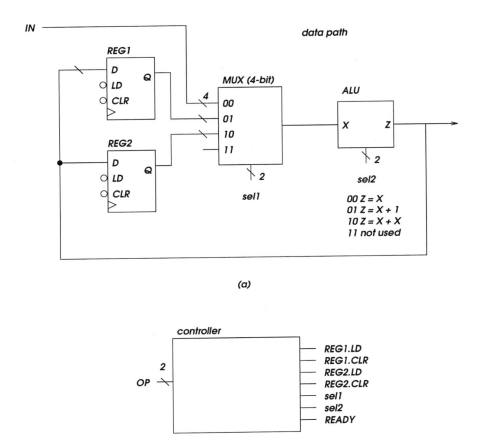

Figure X9.1 (a) Data Path (b) Controller.

The state machine has two input signals, LEFT and RIGHT, that indicate the driver's request for a left turn or a right turn. It also has an emergency-flasher input, HAZ, that requests the tail lights to be operated in hazard mode – all six lights flashing on and off.

9. Shown in Figure X9.1(a) are the circuit elements for a digital circuit that functions as follows:

    If OP = 00, then REG2 <- REG1;

>   followed by REG1 <- IN + 1.
>   If OP = 01, then REG2 <- 0; REG1 <- REG1 + 1.
>   If OP = 10, then REG1 <- IN; REG2 <- IN;
>   followed by REG2 <- REG2 + 1.
>   If OP = 11, then REG1 <- REG1 * 2; REG2 <- REG2 + 2.

   The block diagram for the controller is given in Figure X9.1(b). Optimize the controller by using the least number of states. Describe the whole design in VHDL.

10. You are to design a fuzzy logic elevator controller. There is an elevator in a three-story building. A controller uses the following algorithm to decide which floor the elevator will stay if it is idle. At first the elevator stops at first floor. After someone uses it, the elevator will stay at the most frequently used floor if it is idle. For example, if there is 10 occurrences of entering and leaving the elevator at the third floor, 8 at the second floor and 9 at the first floor. The elevator will stay at the third floor. Here leaving and entering the same person is considered two separate occurrences. Write a VHDL program at FSMD level for the circuit. (**Hint:** When a person enters the elevator, it should notify the controller which floor she/he enters and which floor she/he leaves; for simplicity, consider only one person enters and leaves the elevator each time)

10
MODELING AT THE ALGORITHMIC LEVEL

Different from an RTL or an FSMD level descriptions where a designer has to specify a clock by clock behavior of a circuit, an algorithmic specification looks much more like an ordinary program, consisting of processes containing loops, conditionals, input/output (I/O), and arithmetic and logic expressions. Algorithmic descriptions do not have reference to states and need to be scheduled before they can be mapped to an FSMD description. Table 10.1 shows the difference between algorithmic descriptions and RTL or FSMD descriptions.

	RT or FSMD Level	Algorithmic Level
Description	describing structure	describing algorithm
Behavior	cycle-by-cycle	functional
Design partitions	user specified	auto-generated control, datapath and memory
Operation schedule	input defined	auto-generated
FSM	input defined	auto-generated
Storage elements	input defined	auto-generated
Program length	long	short
Simulation speed	slow	fast

Table 10.1 Comparison of RTL and Algorithmic descriptions

10.1 PROCESS AND ARCHITECTURE

The **process** statement is the main construct in the modeling of sequential behavior of a digital circuit. An algorithmic behavior is expressed in a sequential language using sequential assignments (such as *signal* or *variable* assignments) for data transfer or I/O, control constructs (such as *if* or *case*) for conditional execution, loop statements (such as *for loop* or *while loop*) for iterative sequencing, and *wait* statements for synchronization and handshaking. This type of behavior comes closest to programs written in a standard imperative programming language.

Compared with an RTL description, an algorithmic description may have multiple *wait* statements and allows *while ... loop* statements in a process. In VHDL, restrictions are applied on *wait* statements. Signal edge-detecting attributes (*'event*) have to be included in a *wait* statement in order to assure correct compilation of a VHDL program into a logic circuit. For a process contains *wait* statements, a simulator evaluates it only only at certain discrete points of the simulation time (activated on by signal events). Since signal carriers must maintain its current value unless they are changed by other signal assignments, to make the synthesized circuits meet the VHDL semantic, a dedicated register must be allocated for the signals that are assigned values. Variables, on the other hand, are used for storing intermediate result. They may not be necessarily stored in a register. Furthermore, multiple variables can be stored in one register if there is no conflict (life spans of the variables do not overlap).

An algorithmic process will be synthesized into a circuit containing a datapath, memory, and a controller as shown in Fig. 10.1. The controller issues commands to the datapath and the memory system. The datapath is a point-to-point architecture which consists of registers, multiplexors and functional units. All signal outputs are registered for reasons mentioned previously. The datapath performs operations in accordance with a synthesized schedule. This schedule consists of a set of states and a set of operations in the states. The controller controls the datapath's operations by issuing commands to the data path at the proper times. A synthesizer will translate an algorithmic description into a schedule that may need a minimum hardware for circuit and has the maximum overall performance while meeting the constraints specified by the designers.

Modeling at the Algorithmic Level   193

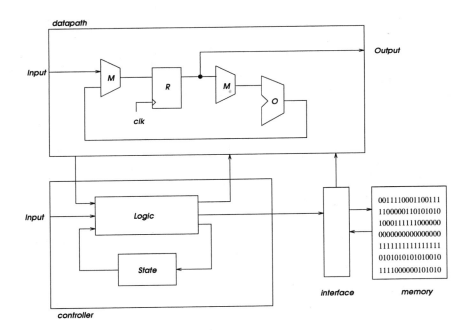

Figure 10.1 Target Architecture.

10.2 WAIT STATEMENTS

In a design description, a process containing **wait** statements implies synchronous logic. A wait statement is used to suspend a process until a positive-edge or negative-edge is detected on a signal where the *signal* being waited on is usually a clock signal. All the storage elements must be triggered by either a leading edge or a falling edge of the clock. Because the scheduling algorithms use the clock cycle time to assign operations into states, each process can only have one clock signal.

Wait statements and signal assignment statements are used by a process to send or receive information from other processes or I/O interfaces. In VHDL synthesis, restrictions are applied on **wait** statements. Signal edge-detecting attributes ('**event**) have to be included in a **wait** statement in order to assure correct compilation of a VHDL program into a logic circuit. There are two groups of wait statements in a synthesizable VHDL design description. The first group includes the following statements:

> **wait on** *signal* **until** *signal = value* ;
> **wait until** *signal*'**event and** *signal = value* ;
> **wait until not** *signal*'**stable and** *signal = value* ;
> **wait until** *signal = value* ;

These four statements have the same meaning. They suspend a process until an event occurs to the signal and the signal changes to the value specified.

The second group includes the following statements:

> **wait until** *signal*'**event and** *signal = value* **and** *expression* ;
> **wait until not** *signal*'**stable and** *signal = value* **and** *expression* ;
> **wait on** *signal* **until** *signal = value* **and** *expression* ;

These three statements suspend a process until all the following three conditions become true (1) a signal event occurs, (2) the signal changes to the value specified, and (3) the expression is true.

Inferred flip-flops will be created for signals that are assigned values within a process containing **wait** statements. For example, the following VHDL description describes the behavior of a D flip-flop. It waits until the leading edge of

Modeling at the Algorithmic Level 195

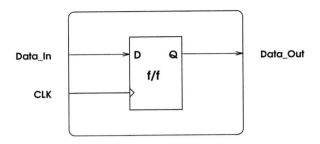

Figure 10.2 A D-type flip-flop.

signal CLK and assigns the value of Data_In to Data_Out. This is the behavior of a D flip-flop as shown in Fig. 10.2. Note that this is a degenerate case of the target architecture in which the whole process is synthesized as a datapath which contains a single flip-flop.

```
signal CLK, Data_In, Data_Out :  BIT ;
. . .
process
begin
   wait until (CLK'event and CLK = '1') ;
   Data_Out <= Data_In ;
end process ;
```

When the clock signal is declared as STD_LOGIC type, it is more general and correct to model the leading edge or the falling edge by the function *rising_edge*() or *falling_edge*(). These functions are defined in both package STD_LOGIC_1164 (for STD_ULOGIC signals) and in package NUMERIC_BIT (for BIT signals). For example :

```
signal CLK, Data_In, Data_Out :  STD_LOGIC ;
. . .
process
begin
   wait until rising_edge(CLK) ;
   Data_Out <= Data_In ;
end process ;
```

10.3 SYNCHRONOUS RESET

When a synchronous reset signal is asserted(active-high or active low), the synthesized design will be forced to the state that is the equivalent of the beginning of the process and some of the registers will be cleared or preset to certain values. In VHDL, a synchronous reset signal should be tested after each wait statement. Since VHDL does not provide a **go to** statement, one way to jump to a synchronous event is by using **loop** and conditional **exit** statements. A global synchronous reset can be accomplished by following every **wait** statement with a conditional exit from the outermost loop.

There are two ways to model a synchronous reset. In the first method, an outermost labeled loop that encloses every statement in the process is defined. Then, every wait statement in the process must be followed by a conditional exit from the outermost loop. Finally, the synchronous reset events are described immediately following the outermost loop. For example, the following program shows a flip-flop with a synchronous input.

```
signal CLK, S_RESET, Data_In, Data_Out:  BIT ;
. . .
process
begin
   S_RESET_LOOP:
     - - normal operations
     loop
        wait until (CLK'event and CLK = '1') ;
        exit S_RESET_LOOP when S_RESET = '1' ;
        Data_Out <= Data_In ;
     end loop S_RESET_LOOP;
     - - synchronous event
     Data_Out <= '0' ;
end process ;
```

The scheduling algorithm of the MEBS system will translate the above algorithmic process into the following RTL process. The schematic of the synthesized circuit is shown in Fig. 10.3(a) and the timing diagram is shown in Fig. 10.3(b).

```
process (CLK)
begin
   if (CLK'event and CLK = '1') then
```

Modeling at the Algorithmic Level

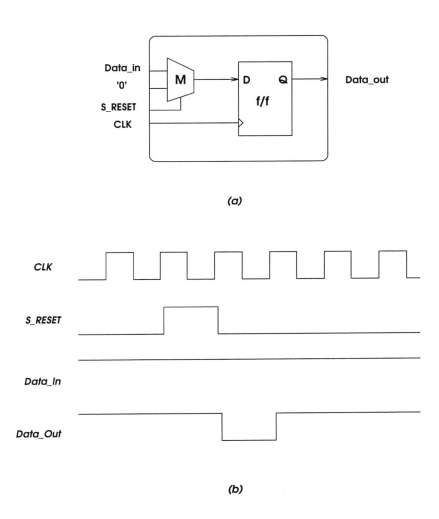

Figure 10.3 Type 1 Synchronous Reset (a) Synthesized Circuit (b) Timing Diagram.

```
        if (S_RESET = '1') then
          Data_Out <= '0' ;
        else
          Data_Out <= Data_In ;
        end if ;
      end if ;
    end PROCESS ;
```

Another way of specifying synchronous reset is to define a global loop which encloses all the statements of a process. The synchronous reset events are described within the loop followed by nornal operations. For each wait statement, a conditional exit to the outermost loop is inserted. For example,

```
    signal CLK, S_RESET, Data_In, Data_Out :  BIT ;
    . . .
    process
    begin
       S_RESET_LOOP:
         -- reset state
         loop
           wait until (CLK'event and CLK = '1') ;
           exit S_RESET_LOOP when S_RESET = '1' ;
           Data_Out <= '0' ;
           main:
             -- normal operations
             loop
               wait until (CLK'event and CLK = '1') ;
               exit S_RESET_LOOP when S_RESET = '1' ;
               Data_Out <= Data_In ;
             end loop ;
         end loop S_RESET_LOOP;
    end process ;
```

The scheduling algorithm of the MEBS system will translate the above algorithmic process into the following RTL process. Note that because in the original specification, there are two wait statements. It will be scheduled into two states. We will need a flip-flop to keep the status of the machine and a

Modeling at the Algorithmic Level

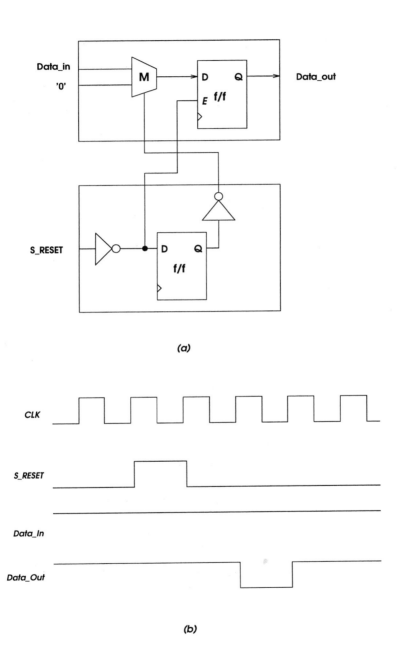

Figure 10.4 Type 2 Synchronous Reset (a) Synthesized Circuit (b) Timing Diagram.

flip-flop to store output signal Data_Out. The description will be synthesized into one in Fig. 10.4(a).

In this description, the synchronous reset event will be executed after the S_RESET signal becomes '0' and because the output signal is registered, the timing will be delayed for one clock cycle. Fig. 10.4(b) shows the timing diagram of the circuit.

```
process (CLK)
   type S_Type is (ST0 , ST1 );
   variable STATE : S_Type := ST0 ;
begin
   if (CLK'event and CLK = '1') then
      case STATE is
         when ST0 =>
            if (S_RESET = '1') then
               STATE := ST0 ;
            else
               Data_Out <= '0';
               STATE := ST1 ;
            end if ;
         when ST1 =>
            if (S_RESET = '1') then
               STATE := ST0 ;
            else
               Data_Out <= Data_In;
               STATE := ST1 ;
            end if ;
      end case ;
   end if ;
end process ;
```

10.4 ASYNCHRONOUS RESET

With restrictions on the format of wait statements in an algorithmic process, assignments of signals have to occur on the edge of a clock signal (synchronous

Modeling at the Algorithmic Level

logic). Asynchronous events become very difficult to model in VHDL. What will happen if an asynchronous signal is asserted in the middle of a circuit operation? The circuit will be reset to its initial state immediately and some of the registers in the circuit will be preset to their initial values. In behavior synthesis, the initial state is the first state that the scheduling algorithm partitioned and the initial values of the registers are the initial values of the variables or signals which are assigned to the registers. Since we know exactly what will be done when an asynchronous reset signal becomes true, we can use a directive to specify the asynchronous reset signal. A synthesis system, after detecting the directive, will connect an asynchronous reset signal to the relevant registers in the circuit.

The asynchronous set/reset signal is specified in the MEBS system by adding an asynchronous set/reset directive in an algorithmic description. The directive - - mebs reset_low means the asynchronous signal is active low and the directive - - mebs reset_high means the asynchronous signal is active high. When an asynchronous set/reset is set, the flip-flops which store signals and variables with initial values will be set to their initial values. Furthermore, the circuit will reset to its initial state.

The following VHDL example shows a D-type flip-flop with an active low asynchronous reset input. The schematic is shown in Fig. 10.5. The directive specifies that there is an active low asynchronous input to the circuit. In the real circuit, an active low asynchronous RESET siganl will force the circuit to transfer to the initial state which is the equivalent of the beginning of the process. Noted that an asynchronous input is one bit wide and should not be referenced elsewhere in the design.

```
entity FF2 is
   port( CLK, RESET, Data_In :  in  BIT ;
         Data_Out             :  out BIT := '0' ) ;
end FF2 ;
architecture BHV of FF2 is
begin
   process
   -- mebs reset_low RESET
   begin
      wait until (CLK'event and CLK = '1') ;
      Data_Out <= Data_In ;
   end process ;
```

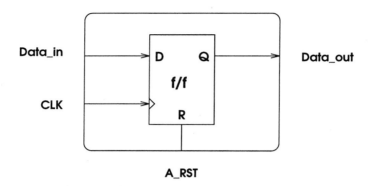

Figure 10.5 A D-type flip-flop with asynchronous reset.

```
    end BHV ;
```

The scheduling algorithm will translate the above algorithmic process into the following RTL process.

```
    process(RESET, CLK)
    begin
       if (RESET = '0') then
          Data_Out <= '0' ;
       elsif (CLK'event and CLK = '1') then
          Data_Out <= Data_In ;
       end if ;
    end process ;
```

10.5 REGISTERS AND COUNTERS

We can describe a four-bit register with an asynchronous set/reset value of "1100" using the VHDL program in the following example. Fig. 10.6 shows the schematic of a four bit register. The active high asynchronous reset signal will set the first two flip-flops to '1' and reset the third and fourth flip-flops to '0'.

```
    entity REG4 is
       port ( CK     : in  BIT ;
```

Modeling at the Algorithmic Level

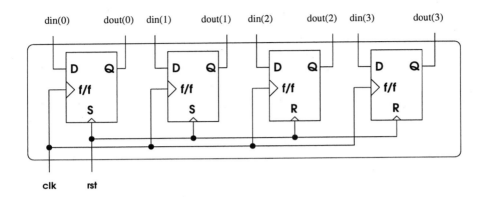

Figure 10.6 A 4-bit register.

```
            RST   : in  BIT ;
            Din   : in  BIT_VECTOR(3 downto 0) ;
            Dout  : out BIT_VECTOR(3 downto 0) := "1100" ) ;
end REG4 ;
architecture BHV of REG4 is
   -- mebs reset_high RST
begin
   process
   begin
      wait until (CK'event and CK = '1') ;
      Dout <= Din ;
   end process ;
end BHV ;
```

The following describes a 4-bit counter with synchronous reset, parallel-in, parallel-out, count-up and count-down. When the synchronous reset signal is asserted, the counter will be set to an initial value of 5.

```
package TYPES is
   constant SIZE : integer := 16 ;
   subtype INT4 is INTEGER range 0 to SIZE - 1 ;
end TYPES ;
```

```
use WORK.TYPES.all ;
entity COUNTER is
   PORT ( CLOCK     : in  BIT ;
          S_RESET   : in  BIT ;
          CountIn   : in  INT4 ;
          Up, Load  : in  BIT ;
          CountOut  : buffer INT4 ) ;
end COUNTER ;
architecture BHV of COUNTER is
begin
   process
   begin
      S_RESET_LOOP:
        loop
           wait until CLOCK'event and CLOCK = '1' ;
           exit S_RESET_LOOP when S_RESET = '1' ;
           if (Load = '1') then
              CountOut <= CountIn ;
           else
              if (Up = '1') then
                 CountOut <= (CountOut + 1) mod SIZE ;
              else
                 CountOut <= (CountOut - 1) mod SIZE ;
              end if ;
           end if ;
        end loop S_RESET_LOOP ;
        CountOut <= 5 ;
   end process ;
end BHV ;
```

10.6 SIMPLE SEQUENTIAL CIRCUITS

During the synthesis process, the statements in a process usually infer certain hardware mapping. For example, the control constructs such as the **if** state-

Modeling at the Algorithmic Level

ments or **case** statements normally infer multiplexers, signal assignments infer assigning value to registers, and expressions infer computation of signals or variables through function units.

The allocation algorithm may perform sharing by assigning several variables to the same register and several operations to the same functional if there is no execution conflict. Due to the sharing, multiplexers may be introduced to gate the correct input to the registers or function units.

In the following program, the first statement is a **wait** statement. The sensitive list of the process consists of signal CLK. It means the process will be evaluated on every leading edge of the signal CLK. If the condition "(SEL = '1')" is true, the value of "(X or Y)" will be assigned to signal Data_Out. If not, the signal will keep the previous value. Since the assignment will happen only at the leading edge of signal CLK, it means the value will be assigned into a D flip-flop. Fig. 10.7 shows the synthesized circuit which consists of a flip-flop, a multiplexer and an OR-gate.

```
entity MUX_REG is
   port( CLK, SEL : BIT ;
         X, Y     : BIT ;
         Data_Out :  out BIT ) ;
end MUX_REG ;
architecture BHV of MUX_REG is
begin
   process
   begin
      wait until (CLK'event and CLK = '1') ;
      if (SEL = '1') then
         Data_Out <= X or Y ;
      end if ;
   end process ;
end BHV ;
```

10.7 ALGORITHMS

The sequential statements in VHDL can be used to describe any algorithm. In the modeling of an algorithm, the signal assignment statements are used as a

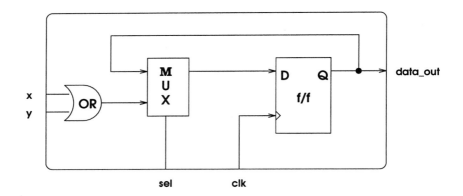

Figure 10.7 A gate level schematic for the simple sequential circuit.

vehicle to communicate with the outside world and the **wait** statements are used to receive data from outside. In the following example, we show a very simple algorithm which is used to calculate the greatest common divisor (GCD) of two eight-bit numbers. An algorithmic description of the GCD calculator algorithm is shown in Fig. 10.8.

The following example shows a VHDL description of the GCD calculator. When the READY signal is high at a leading edge of the CLOCK signal, the two eight-bit signals are moved into local variables. A **while ... loop** is used to compute the greatest common divider (GCD) of the two numbers. When the computation is over, the result is moved to the output port OU.

```
entity GCD is
   port ( CLOCK : in  BIT ;
          RESET : in  BIT ;
          XI,YI : in  INTEGER range 0 to 255 ;
          READY : in  BIT ;
          OU    : out INTEGER range 0 to 255 := 0 ) ;
end GCD ;
architecture BHV of GCD is
   -- mebs reset_high RESET
begin
  Main : process
```

DECLARATION	X(7:0), Y(7:0)
INPUT:	X <- Xi; Y <- Yi;

CALCULATION:	while(X /= Y) loop
	   if (X < Y) then Y <- Y - X;
	   else X <- X - Y;
	end loop;

OUTPUT:	ou <- X;

Figure 10.8 Algorithmic description of an eight-bit GCD calculator.

```
   variable X, Y : INTEGER range 0 to 255 ;
 begin
   wait until(CLOCK'event and CLOCK = '1' and READY = '1');
   X := XI ;
   Y := YI ;
   while (X /= Y) loop
     wait until(CLOCK'event and CLOCK = '1');
     if (X < Y) then
       Y := Y - X ;
     else
       X := X - Y ;
     end if ;
   end loop ;
   OU <= X ;
 end process Main ;
end BHV ;
```

Note that we have to insert a "wait until(CLOCK'event and CLOCK = '1');" statement in the loop to ensure the behavior of the circuit matches that of the synthesized circuit. Without the wait statement, the simulator will execute the loop in zero time which means the result will come out immediately after the input is ready, while it may take several clock cycles for the real circuitry to produce a result.

DECLARATION:	A(7:0), M(7:0), Q(7:0), COUNT(2:0), F
INPUT:	M <- m1;
	Q <- m2;
ADD:	A(7:0) <- A(7:0) + M(7:0) *Q(0);
	F <- M(7) and Q(0) or F;
RIGHTSHIFT:	A(7) <- F;
	A(6:0)&Q <- A &Q(7:1);
TEST:	if (COUNT = 6) then go to CORRECTION;
	COUNT <- COUNT + 1;
	go to ADD;
CORRECTION:	A(7:0) <- A(7:0) - M(7:0) *Q(0);
OUTPUT:	result <- F & A & Q(7:1);

Figure 10.9 Algorithmic Description of an eight-bit two's-complement multiplier.

The next example is to multiply two eight-bit two's-complement numbers, $m_1(7:0)$ and $m_2(7:0)$, and put the result in $result(15:0)$ using an shift-and-add multiplication algorithm.. An algorithmic description is shown in Fig. 10.9.

The algorithm can be easily translated into the following VHDL description.

```
library IEEE ;
use IEEE.NUMERIC_BIT.all ;
package BOOTH_TYPES is
  constant BOOTH_SIZE         : INTEGER := 8 ;
  subtype BOOTH_COUNTER_TYPE is INTEGER range 0 to (BOOTH_SIZE - 1);
  subtype BOOTH_IN_RANGE is INTEGER range (booth_size - 1) downto 0;
  subtype BOOTH_IN_TYPE is UNSIGNED(BOOTH_IN_RANGE) ;
  subtype BOOTH_OUT_RANGE is INTEGER range (2*booth_size-1) downto 0;
  subtype BOOTH_OUT_TYPE is UNSIGNED(BOOTH_OUT_RANGE) ;
end BOOTH_TYPES ;
use WORK.BOOTH_TYPES.all ;
entity MULT is
```

```
   port ( CLOCK   : in  BIT ;
          RESET   : in  BIT ;
          In_Rdy  : in  BIT ;
          Out_Rdy : out BIT := '0' ;
          M1      : in  BOOTH_IN_TYPE ;
          M2      : in  BOOTH_IN_TYPE ;
          Qout    : out BOOTH_OUT_TYPE ) ;
   constant SIZE  : INTEGER := BOOTH_SIZE - 1;
end MULT ;
architecture BHV of MULT is
   -- mebs reset_high RESET
begin
   MUL :
   process
      variable COUNTER : BOOTH_COUNTER_TYPE := 0 ;
      variable A : BOOTH_IN_TYPE := BOOTH_IN_TYPE'(others => '0') ;
      variable M,Q     : BOOTH_IN_TYPE ;
      variable F       : BIT := '0' ;
   begin
      -- input
      wait on CLOCK until CLOCK = '1' and In_Rdy = '1' ;
      M := M1 ;
      Q := M2 ;
      loop
         wait until CLOCK'even and CLOCK='1' ;
         -- add
         if (Q(0) = '1') then
            A := A + M ;
         end if ;
         F := (M(SIZE) and Q(0)) or F ;
         -- right shift
         Q := A(0) & Q(SIZE downto 1) ;
         A := F & A(SIZE downto 1) ;
         -- test
         exit when (COUNTER = (BOOTH_SIZE - 2)) ;
```

```
              COUNTER := COUNTER + 1 ;
          end loop ;
          -- correction
          if (Q(0) = '1') then
              A := A - M ;
          end if ;
          -- output
          Out_Rdy <= '1' ;
          Qout <= F & A & Q((BOOTH_SIZE - 1) downto 1) ;
          wait on CLOCK until CLOCK = '1' and In_Rdy = '0' ;
          A := BOOTH_IN_TYPE'(others => '0') ;
          COUNTER := 0 ;
          F := '0' ;
          Out_Rdy <= '0' ;
      end process ;
end BHV ;
```

Another interesting and widely used scheme for two's-complement multiplication was designed by Andrew D. Booth. Unlike the preceding algorithm, it scans the multiplier from right to left and using the value of the current multiplier bit to determine which of the following operations to perform: add to the multiplicand, subtract the multiplicand, or perform no operation. In Booth's approach two adjacent bits are examined in each step. If it is 01, then the multiplicand is added to the partial product, while if it is 10, the multiplicand is subtracted from the partial product. If it is 00 or 11, then neither addition nor subtraction is performed; only the subsequent right shift of the partial product takes place. By doing so, the Booth's algorithm can effectively skip over runs of 1s and runs of 0s that it encounters. Fig. 10.10 shows an algorithmic description of an 8-bit (modified) Booth's multiplier.

The following shows a VHDL description for the modified Booth's multiplier.

```
architecture MODIFIED of BOOTH is
    -- mebs reset_high RESET
begin
    MUL :
    process
        variable COUNTER : BOOTH_COUNTER_TYPE := 0 ;
```

```
DECLARATION: A(7:0), M(7:0), qq, Q(7:0), COUNT(2:0), F
INPUT:       M <- m1;
             Q <- m2;
             qq <- m2(7);
ZEROTEST:    if (M = 0) or (Q = 0) then go to OUTPUT;
LOOP:        if (Q(1:0) = 01) then
                 if (F = 0) then A <- A + M;
             elsif (Q(1:0) = 11) then
                 if (F = 0) then A <- A - M; F <- 1;
             elsif (Q(1:0) = 00) then
                 if (F = 1) then A <- A + M; F <- 0;
             else
                 if (F = 1) then A <- A - M;
RIGHTSHIFT:  Q <- qq & Q(7:1);
             qq <- A(0);
             A <- ((not M(7) and F) or (M(7) and A(7))) & A(7:1);
TEST:        if (COUNT = 7) then go to OUTPUT;
INCREMENT:   COUNT <- COUNT + 1;
             go to LOOP;
OUTPUT:      result <- A & qq & Q97:1);
```

Figure 10.10 Algorithmic description of an 8-bit (modified) Booth multiplier.

```vhdl
      variable A : BOOTH_IN_TYPE := BOOTH_IN_TYPE'(others => '0') ;
      variable M,Q     : BOOTH_IN_TYPE ;
      variable F       : BIT := '0' ;
      variable QQ      : BIT ;
   begin
      -- input
      wait on CLOCK until CLOCK = '1' and In_Rdy = '1' ;
      M := M1 ;
      Q := M2 ;
      QQ := Q(SIZE) ;
      -- zero test
      if not (M = 0 or Q = 0) THEN
         loop
            wait on CLOCK until CLOCK = '1' ;
            case Q(1 downto 0) is
               when "01" =>
                  if (F = '0') then
                     A := A + M ;
                  end if ;
               when "11" =>
                  if (F = '0') then
                     A := A - M ;
                     F := '1' ;
                  end if ;
               when "00" =>
                  if (F = '1') then
                     A := A + M ;
                     F := '0' ;
                  end if ;
               when "10" =>
                  if (F = '1') then
                     A := A - M ;
                  end if ;
            end case ;
            -- right shift
```

```
                Q := QQ & Q(SIZE downto 1) ;
                QQ := A(0) ;
                A := ((not M(SIZE) and F) or (M(SIZE) and A(SIZE))) &
                     A(SIZE downto 1) ;
                -- test
                exit when (COUNTER = SIZE) ;
                COUNTER := COUNTER + 1 ;
            end loop ;
        end if ;
        -- outout
        Out_Rdy <= '1' ;
        Qout <= A & QQ & Q(SIZE downto 1) ;
        wait on CLOCK until CLOCK = '1' and In_Rdy = '0' ;
        A := BOOTH_IN_TYPE'(others => '0') ;
        COUNTER := 0 ;
        F := '0' ;
        Out_Rdy <= '0' ;
    end process ;
end MODIFIED ;
```

10.8 PROCESS COMMUNICATION

A digital system can be described as a set of processes which communicate with each other using signals. In VHDL, variables must be declared and used inside a process. They are only "visible" within a process and cannot be used for process communication. Signals are used for process communication. They are declared within the same architecture so that they will be "visible" and available to all the processes within the architecture. Fig. 10.11 (a) shows a diagram illustrating how two processes communicate with each other. Process B suspends the execution until the condition in the **wait** clause becomes true. Process A activates process B by setting the **req** signal to one. Similarly, process B activates process A by setting the **ack** signal to one.

A system consisting of multiple communicating processes will be synthesized into a system consisting of a set of interconnected components. Each process is implemented as a component consisting of a data path, control unit and

214   CHAPTER 10

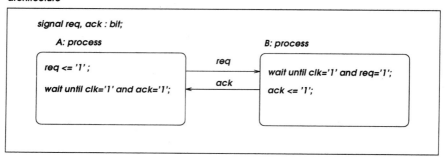

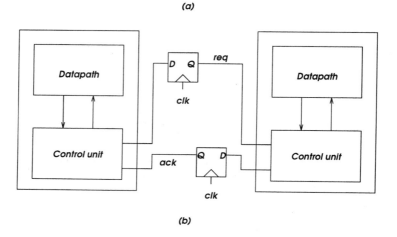

Figure 10.11 Target architecture (a) Process communication (b) FSMD architecture

memory. Because a signal will maintain its previous value until a transaction is scheduled to the signal, it has to be stored in a flip-flop. In other words, a signal assignment statement in an algorithmic process infers the signal to be assigned will be latched into a flip-flop. Figure 10.11 (b) shows such a block diagram for the two communicating processes.

The timing relationship between the signals among the communicating processes is called the communication protocol. In the following two subsections, examples are used to illustrate two commonly used communication protocols.

10.8.1 Two Way Handshaking Communication

Two-way handshaking communication uses a control line for the sender to communicate with the receiver and another control line for the receiver to communicate with the sender. The sender interrupts the receiver by setting the request signal to 1. It suspends the execution until the acknowledge signal is set to 1 by the receiver process, at which time it resets the request signal back to 0. The receiver waits until it sees the request signal changes to 1. It then sets the acknowledge signal to 1 and proceeds to the operations to be executed. In a two-way handshaking communication, the sender and receiver exchange data by exchanging control signals. In such a communication scheme, the data is transmitted when handshaking is completed.

Two-way handshake protocols are typically used in a communication where there is no fixed response time between the sender and the receiver. The scheduling algorithm may partition the statements between two wait statements into several states. The following shows an example of a two-way handshaking communication. Process A first puts the data on the DATA line and issues a READY signal to process B. Then, it suspends the execution until it sees an acknowledgement from process B. After reactivation, it resets the READY signal and waits until process B resets the ACK signal. After process B resets the ACK signal, process A can continue to transmit the next data. Fig. 10.12 shows the simulation waveform of the the handshaking communication.

```
entity HANDSHAKE is
    port ( CLK      : in bit ;
           Data_In  : in INTEGER ;
           Data_Out : out INTEGER) ;
end HANDSHAKE ;
```

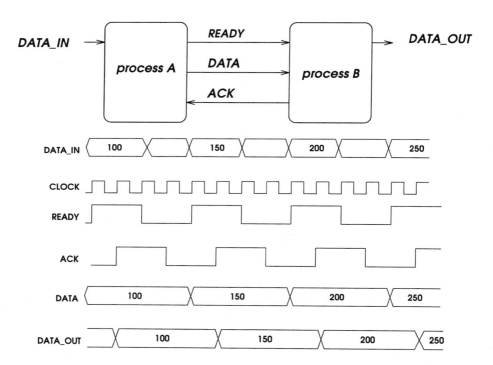

Figure 10.12 Two way handshaking communication.

```
architecture PROTOCOL of HANDSHAKE is
   signal Ready,   ACK : bit := 0 ;
   signal Data           : INTEGER ;
begin
   A: process
   begin
      wait until CLK'event and CLK = '1' and ACK = '0' ;
      Data <= Data_In ;
      Ready <= '1' ;
      wait until CLK'event and CLK = '1' and ACK = '1' ;
      Ready <= '0' ;
   end process ;
   B: process
   begin
      wait until CLK'event and clk = '1' and Ready = '1' ;
      Data_Out <= Data ;
      ACK <= '1' ;
      wait until CLK'event and CLK = '1' and Ready = '0' ;
      ACK <= '0' ;
   end process ;
end PROTOCOL ;
```

10.8.2 One Way Handshaking Communication

Communication by two way handshaking is very timing consuming especially when there is a block of data which has to be transmitted to the receiver over a data line. Each communication may take several clock cycles to complete using the handshaking protocol. If the receiver knows exactly when the sender will transmit the data, then communication can be made much faster.

This scheme is commonly used in a one way communication where the receiver suspends the execution and waits for a ready signal from the sender. When the sender is ready to transmit data, he sends a ready signal to the receiver. After that, the sender will transmit the data to the receiver based on a predetermined rate. The communication completes after a block of data has been transmitted.

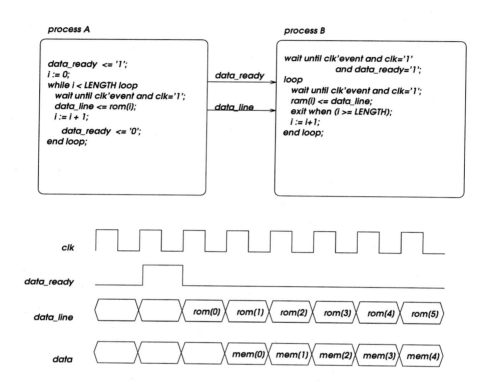

Figure 10.13 Source oriented data communication.

Because a predetermined rate is set for such a communication scheme, it is the responsibility of the designer to specify the cycle-by-cycle based behavior for both the sender process and the receiver process. In such a scheme, the scheduler is not allowed to cut the statements between two successive wait statements into several states because it becomes impossible to deliver the data at a predetermined rate. Therefore, the scheduler should provide an option to preserve the cycle-by-cycle behavior specified by a designer.

In the following example, process A has a 4×16 ROM and process B has a 4×16 RAM. Process A will transmit the data in the ROM to process B using this communicating protocol. Process A first sends a Data_Ready signal to process B, and then starts transmitting the data to process B one word per clock cycle. After being activated by the Data_Ready signal, the receiver will move the data from the data line into its buffer; one word per clock cycle. Fig. 10.13 shows the program segment, and timing for such a communication scheme.

```
entity ONE_WAY is
    port ( CLK       : in  BIT ;
           Data_In   : in  INTEGER range 0 to 255 ;
           Data_Out  : out INTEGER range 0 to 255 ) ;
end ONE_WAY ;
architecture PROTOCOL of ONE_WAY is
    subtype INT4 is INTEGER range 0 to 15 ;
    subtype INT8 is INTEGER range 0 to 255 ;
    type MEMORY is array(INT4) of INT8 ;
    signal Data_Ready  : BIT ;
    signal Data_Line   : INT8 ;
begin
    A : process
        variable I    : INT4 ;
        constant ROM  : MEMORY :=
                (16#4B#, 16#12#, 16#56#, 16#83#,
                 16#20#, 16#45#, 16#1#,  16#B1#,
                 16#41#, 16#72#, 16#1A#, 16#D3#,
                 16#22#, 16#15#, 16#C5#, 16#99# ) ;
    begin
       Data_Ready <= '1' ;
       I := 0 ;
       loop
          wait until CLK'event and CLK = '1' ;
          Data_Line <= ROM(I) ;
          exit when (I = INT4'high) ;
          I := I + 1 ;
          Data_Ready <= '0' ;
       end loop ;
    end process ;
    B : process
        variable I   : INT4;
        variable MEM : MEMORY ;
```

```
      begin
        wait until CLK'event and CLK = '1' and Data_Ready = '1' ;
        I := 0 ;
        loop
            wait until CLK'event and CLK = '1' ;
            MEM(I) := Data_Line ;
            exit when (I = INT4'high) ;
            I := I + 1 ;
        end loop ;
      end process ;
    end PROTOCOL ;
```

10.9 SUMMARY

1. An algorithmic specification looks much more like an ordinary program, consisting of processes containing loops, conditionals, input/output (I/O), and arithmetic and logic expressions.

2. The **process** statement is the main construct in the modeling of the sequential behavior of a digital circuit. A **wait** statement is used to suspend a process until a positive-edge or negative-edge is detected on a signal in an algorithmic level description.

3. In a behavior modeling, we have to insert a "wait until(CLOCK'event and CLOCK = '1');" statement in the loop to ensure the behavior of the circuit matches that of the synthesized circuit.

4. Two-way handshaking communication uses a control line for the sender to communicate with the receiver and another control line for the receiver to communicate with the sender. It is typically used in a communication where there is no fixed response time between the sender and the receiver.

5. One way communication is used when there is a predetermined rate between sender and receiver.

Exercises

1. Write a behavior description for a divide-by-n circuit where n is passed to the circuit through a generic parameter. The input is I and the output is O. The circuit has a synchronous reset input that resets the circuit to its initial state.

2. An 8-bit parallel-to-serial converter circuit is to be designed. The circuit remains in an idle state as long as the START input is false. But when this START input becomes true, the 8-bit data BYTE is loaded into the shift register and the right-shifting of the data begins. After the 8 bits are shifted out, the circuit returns to the idle state. Write a design description for the circuit at the algorithmic level.

3. Figure X10.1 shows a block diagram which illustrates the handshake between module A and module B. The function of module A is to first issue a signal, r, to module B indicating that an access request signal has been made on one of two lines REQX or REQY, but not on both. Then, if the system acknowledges receipt of the request by sending back a signal ACK to the module A while the request is active, module A will issue an access request (wither REQX or REQY) through signals X, or Y, respectively. At most one of the signals will be active at a time. Write an algorithmic design description for module A.

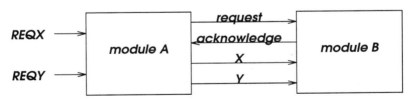

Figure X10.1

4. Convert the algorithmic design description in the previous problem into a design description at FSMD level.

5. Figure X10.2 shows a flowchart describing an algorithm for multiplication that is used in some low-speed computer systems. The algorithm is implemented by three up-down counters, CQ, CM, and CP, which store the multiplier, multiplicand, and product, respectively, and the product CP is formed by incrementing the counter CP a total of CP times. Write a design description for the multiplier.

```
INPUT:      Q <- multiplier ;
            CQ <- multiplier ;
            CM <- multiplicand ;
            CP <- 0 ;

ZEROTEST:   if (CM=0) or (CQ=0) then goto OUTPUT

LOOP:       CQ <- CQ -1 ;
            CP <- CP + 1 ;
            if (CQ /= 0) then go to LOOP;

            CM <- CM -1 ;
            CQ <- Q ;
            if (CM /= 0) then go to LOOP;

OUTPUT:     result <- CP ;
```

Fig. X10.2

6. Consider the design of a multiplier that implements the multiplication algorithm of Figure X10.2. The numbers to be multiplied are four-digit integers in sign-magnitude BCD code. For example, the number 301 is represented by the bit sequence

    0   0000   0011   0000   0001

    CQ, CM, and CP are to be constructed from modulo-10 up-down counters with parallel input-output capability. Write a design description for a design using this multiplication algorithm.

7. Draw a flowchart of a counting algorithm similar to that of Figure X10.2 which performs integer division. Write a VHDL design description for the design.

8. Figure X10.3 shows a flowchart for a typical pencil-and-paper method for the division of unsigned numbers. Convert the design into an algorithmic design description.

9. Figure X10.4 presents a non-restoring division algorithm for unsigned integers. The divisor V and quotient Q are n bits long, while the dividend D is $2n - 1$ bits long, which the maximum length of the product of two n-bit integers. The flip-flop S is appended to the accumulator A to record the sign of the result of an addition or subtraction operation and to determine

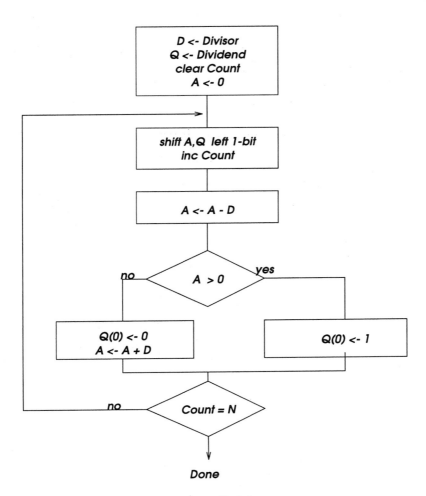

Figure X10.3

PORT:	D(13:0), V(7:0), QUO(7:0), REM(7:0)
DECLARATION:	S, A(7:0), Q(7:0), M(7:0), Count(2:0)
BEGIN:	Count := 0; S := 0;
INPUT:	A := D(13:8); Q := D(7:0); M := V;
SUBTRACT:	S.A := S.A - M; /* S is the sign bit */

TEST:	if S=0 then begin
	  Q(0) := '1';
	  if Count = 7 then go to CORRECTION;
	  else begin
	    Count := Count + 1;
	    S.A.Q(7:1) := A.Q;
	    end;
	  S.A := S.A - M;
	  go to TEST;
	  end;
	else begin /* S = 1 */
	  Q(0) := '0';
	  if Count = 7 then go to CORRECTION;
	  else begin
	    Count := Count + 1;
	    S.A.Q(7:1) := A.Q;
	    end;
	  S.A := S.A + M;
	  go to TEST;
	end;

CORRECTION:	if S=1 then S.A := S.A + M;
OUTPUT:	QUO <= Q; REM <= A;

Figure X10.4

the quotient bit. Each new quotient bit is placed in Q(n-1), and the final values of the quotient Q and the remainder R are in the Q and A registers, respectively. Write an algorithmic design description for the design.

10. You are to design a fuzzy logic elevator controller. There is an elevator in a three-story building. A controller uses the following algorithm to decide which floor the elevator will stay if it is idle. At first the elevator stops at first floor. After someone uses it, the elevator will stay at the most frequently used floor if it is idle. For example, if there is 10 occurrences of entering and leaving the elevator at the 3rd floor, 8 at the second floor and 9 at the first floor. The elevator will stay at the third floor. Here leaving and entering the same person is considered two separate occurrences. Write a VHDL program at the algorithmic level for the circuit controller.

11

MEMORIES

A memory unit stores binary information in groups of bits called *words*. A word in memory is a group of bits that move in and out of storage as a unit. The communication between a memory and its environment is achieved through data input lines, address lines, control lines, and data output lines. Fig. 11.1(a) shows a symbol of 16 × 3-bit random access memory. The two operations that a random access memory can perform are the read, and write operations. The steps to read a data out of memory are as follows:

1. Transfer the address of the desired word to the address lines.

2. The write enable (WE) signal must remain inactive (active LOW in the case of Fig. 11.1(a)) during a read cycle.

3. After a period of time, the data will be ready on the data output lines.

The steps to write a word into memory are as follows:

1. Transfer the address of the desired word to the address lines.

2. Transfer the data bits to the data input lines.

3. Activate the write enable (WE) input (active High in the case of Fig. 11.1(a)).

4. The data bits will be written into the memory after a period of time.

Note that some memory modules may have Chip Enable (CE) control input and use active low Write Enable (WE) signal. A memory module is usually

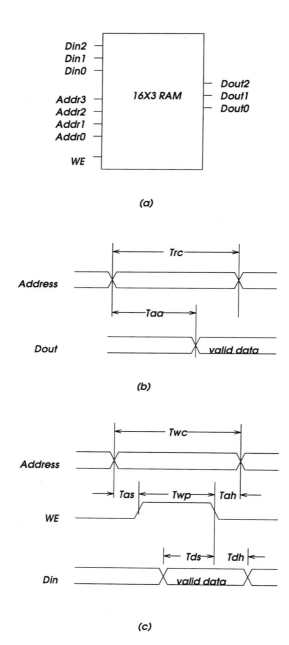

Figure 11.1 (a) A random access memory (b) Memory read timing specification (c) Memory write timing specification

Memories 229

characterized by an *access time* which is the time required to select a memory word to either read or write. Fig. 11.1 (b) and (c) shows the timing specification for a typical memory module. Fig. 11.1(b) shows the timing specification to read a memory module and Fig. 11.1(c) shows the timing specification to write a data into a memory. Here T_{rc} and T_{wc} are the read cycle time and write cycle time, respectively. T_{aa} is the memory access time which is the time period needed for a valid data to appear on the data line after the address is stable. To write a data into a memory, most of the memory modules require the following:

- After the *Address* lines become stable, we must wait at leat T_{as} to raise WE to '1'.

- The WE signal has to keep stable for a minimum time period denoted by T_{wp}.

- After the WE changes back to 0, the *Address* must remain the same value for a time period no shorter than T_{ah}.

- The data must be present to the Din lines at least T_{ds} (set-up time) before WE changes to 0, and it must remain stable for at least T_{dh} (hold time) after WE changes to 0.

11.1 MEMORY READ/WRITE AT THE RT LEVEL

Normally, the external controller which uses the memory module is synchronized with its own clock pulse. The memory, however, does not have internal clock pulses, and its read and write operations are governed by the control signals. The external controller must provide the memory control signals such that the internal clock operations are synchronized with the read and write operations of the memory. Designers are responsible for designing a control logic which generates the signals to the memory modules that meet the specification. This is usually done through an interface circuit between the controller and the memory. Fig. 11.2 shows a diagram representing the relationships among a controller, an interface circuit and a memory module.

In this section, we will use an example to illustrate how to read/write memory at the RT level. Assume the clock cycle rate of the controller is 50 ns and the memory has a 12 ns maximum access time. Let Twp be 8 ns, and Tas and Tah

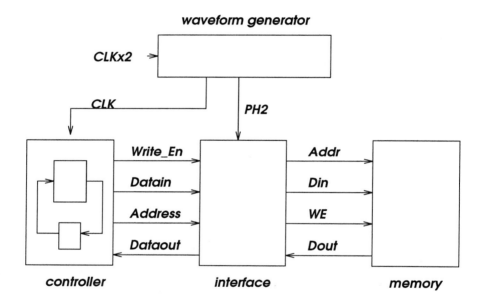

Figure 11.2 A diagram of controller, interface and memory module.

Memories 231

must be at least 2 ns. To access the memory, we need to generate a WE signal which last for at least 8 ns. We can divide the clock into 4 phases and use one phase for the WE signal. To derive a four phase clock from a single phase, the easiest way is to use a clock which is two times the speed of the controller's clock. The program segment in block *Waveform_generator* derives the clock for the controller and generates a PH2 to be used for write enable signal.

To read data from the memory, in the first state we assign the target address to the address lines. The data will be available at the end of the next state. To write data into the memory, in the first state we assign the target address and data to the address lines and data input lines, respectively. At the same time, we assign '1' to the Write_En signal. The Write_En signal will be "and" with the PH2 signal to give the WE signal to the memory. Fig. 11.3 shows the timing of the interface circuit.

```
       package RAM_Test_PKG is
          -- define a 16 X 8 RAM
          subtype Data_Type    is INTEGER range 0 to 255 ;
          subtype Address_Type is INTEGER range 0 to 15 ;
          type RAM_Type is array (Address_Type) of Data_Type ;
          -- 16X8 memory
       end RAM_Test_PKG ;
       use WORK.RAM_Test_PKG.all ;
       entity RAM is
          port ( CLKx2, RESET : in  BIT );
       end RAM ;
       architecture EX OF RAM IS
          -- define a 16 X 8 RAM
          signal Din, Dout, Datain           : Data_Type ;
          signal Addr, Address :   Address_Type ;
          signal WE, Write_En, : BIT := '0' ;
          signal CLK,NEG,PH2      : BIT := '1' ;
          -- mebs reset reset
       begin
       Waveform_generator:
          block
          begin
             CLK_Gen :
```

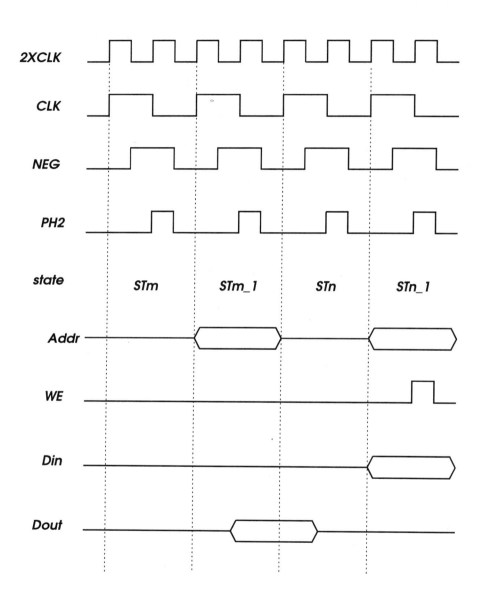

Figure 11.3 Timing of a memory interface.

```
            process (RESET, CLKx2)
            begin
               if (RESET = '1') then
                  CLK <= '1' ;
               elsif (CLKx2'event and CLKx2 = '1') Then
                  CLK <= not CLK;
               end if ;
            end process CLK_Gen ;
         NEG_Gen :
            process (RESET, CLKx2)
            begin
               if (RESET = '1') then
                  NEG <= '1' ;
               elsif (CLKx2'event and CLKx2 = '0') then
                  NEG <= not NEG ;
               end if ;
            end process NEG;
         PH2_Gen :
            PH2 <= not CLK and NEG;
      end block;
   Interface:
      Addr <= Address ;
      WE <= Write_En and PH2 ;
      Datain <= Din ;
      Dataout <= Dout ;
   RAM_module :         -- Memory model for RT level
      process (WE, Addr, Din)
         variable MEM : RAM_Type;
      begin
         case WE is
            when '0' =>  Dout <= MEM(Addr) after 35 ns ;
            when '1' =>  MEM(Addr) := Din ;
         end case ;
      end process RAM_module ;
   Controller:
```

```
process (CLK, RST)
    type STATE_TYPE is (STm,STm_1,STn,STn_1);
    variable State:   STATE_TYPE ;
    variable Target_Addr :   Address_Type ;
    variable Target_Data :   Data_Type ;
    -- mebs state_var State
begin
    if (RST = '1') then
        State := ST0;
    elsif (CLK'event and CLK = '1') then
        case STATE is
            . . .
            when STm =>
                --------- to read data from memory --------
                Address <= Target_Addr ;
                Write_En <= '0' ;
                State := STm_1 ;
            when STm_1 =>
                -- The data in Target_Addr is ready to be used.
                -- ...  <= Dataout ;
            when STn =>
                --------- to write data into memory -------
                Datain <= Target_Data ;
                Address <= Target_Addr ;
                Write_En <= '1' ;
                Next_State := STn_1 ;
            when STn_1 =>
                -- Target_Data is written into memory
            . . .
        end case ;
    end if ;
end process ;
end EX ;
```

Sometimes, the memory cycle time is so short that we may access a memory in half a clock cycle. A good design is to use a clock with faster clock rate and divide it to meet the timing specification. However, if it is not possible, a tricky design (using delay circuits and buffers) can achieve the same purpose. A tricky circuit design can achieve the gap that is needed between the address and the write enable signal, but it may only be used as a last resort.

11.2 MEMORY INFERENCE AT THE ALGORITHMIC LEVEL

An important aspect in behavior synthesis is the ability to synthesize memories and memory interfaces "automatically". It can schedule the memory I/O operations automatically. This provides the designer with a simple coding style for using memories. Arrays are used for modeling linear structures such as registers, RAMs and ROMs. The elements of an array can be addressed using an index variable. The index of an array must be a discrete type (integer type or enumeration type). The example below shows a definition of a 16×10 memory type.

```
subtype INT4    is INTEGER range 0 to 15 ;
subtype INT10   is INTEGER range 0 to 1023 ;
type    Memory  is array ( INT4 ) of INT10 ;
```

An array of variables (constants) is usually synthesized as a RAM (ROM). However, if the number of bits of an element is one, it can be synthesized either as a memory or a register. A directive, **mebs register_array**, has to be added to the array declaration for register declaration. In the following example, A will be synthesized as an 8-bit register, B will be synthesized as an 8×1 memory module, and C will be synthesized as a 6×2 memory module.

```
type COLOR1 is (RED, ORANGE, YELLOW, GREEN, BLUE, PURPLE) ;
type COLOR2 is (GREEN, BLACK, WHITE, YELLOW) ;
-- mebs register_array
type RBYTE is array (NATURAL range <>) of BIT ;
type MBYTE is array (7 downto 0) of BIT_VECTOR(0 downto 0) ;
type MJUNK is array (COLOR1) of COLOR2 ;
variable A : RBYTE(7 downto 0) ;
variable B : MBYTE ;
variable C : MJUNK ;
```

In the following VHDL example, DataIn is a declaration of a 16×8 ROM and DataArray is a declaration of a 16×9 RAM. The program moves the data in the ROM to the RAM, sorts the data in the RAM, and displays the data in sequence on a seven segment display.

```
library IEEE ;
use IEEE.NUMERIC_BIT.all ;
use WORK.P7SEG.all ;
entity RAMROM is
    port ( CLOCK            : in  bit;
           RESET            : in  bit;
           Data_Ready       : in  bit;
           Sorting_Complete : out bit;
           ACK              : out bit;
           SEG1,SEG2        : out UNSIGNED(6 downto 0) ) ;
end RAMROM ;
architecture SORT of RAMROM is
    subtype bit4 is INTEGER range 0 to 15 ;
    subtype bit8 is INTEGER range 0 to 255 ;
    type MemoryROM is array (BIT4) of BIT8 ;
    type MemoryRAM is array (BIT4) of UNSIGNED(7 downto 0) ;
    signal Data_Out :  UNSIGNED(7 downto 0) := "00000000" ;
    -- mebs reset_high RESET
begin
    Sort : process
        constant DataIn            : MemoryROM :=
           (16#4B#, 16#12#, 16#56#, 16#83#, 16#20#, 16#45#, 16#1#, 16#B1#,
            16#41#, 16#72#, 16#1A#, 16#D3#, 16#22#, 16#15#, 16#C5#, 16#99# ) ;
        variable DataArray         : MemoryRAM ;
        variable I                 : BIT4 := 0 ;
        variable Num,Count,Maxi    : BIT4 ;
        variable Max               : UNSIGNED(7 downto 0) ;
    begin
        Sorting_Complete <= '0' ;
        I := 0 ;
```

```
Move_Data: loop
   wait until (CLOCK'event and CLOCK = '1') ;
   DataArray(I) := To_Unsigned(DataIn(I),8) ;
   exit when (I = bit4'high) ;
   I := I + 1 ;
end loop ;
I := 0 ;
Sort_Data : loop
   Max := "00000000" ;
   Count := I ;
   Find : loop
      if (DataArray(Count) > Max) then
         Max  := DataArray(Count) ;
         Maxi := Count ;
      end if ;
      EXIT Find WHEN (Count = BIT4'high) ;
      Count := Count + 1 ;
   end loop Find ;
   DataArray(Maxi) := DataArray(I) ;
   DataArray(I)    := Max ;
   exit Sort when (I = BIT4'high) ;
   i := i + 1 ;
end loop Sort ;
I := 0 ;
Sorting_Complete <= '1' ;
Display_Data : loop
   wait until (CLOCK'event and CLOCK = '1' and
                                Data_Ready = '1') ;
   ACK <= '1' ;
   Data_Out <= DataArray(I) ;
   wait until (CLOCK'event and CLOCK = '1' and
                                Data_Ready = '0') ;
   ACK <= '0' ;
   exit Output_16_Data when (I = BIT4'high) ;
   I := I + 1 ;
```

```
      end loop Output_16_Data ;
    end process ;
    SEG1 <= decode_H4b(Data_Out(7 downto 4)) ;
    SEG2 <= decode_H4b(Data_Out(3 downto 0)) ;
end Sort ;
```

11.3 SUMMARY

1. A memory unit stores binary information in groups of bits called *words*. The communication between a memory and its environment is achieved through data input and output lines, address lines, and control lines that specify the direction of transfer.

2. To access a memory without accidentally destroying the data in the memory, some memory modules require that 1) the WE signals be activated after the signals in the address lines are stable, and 2) the address must remain stable for a short period after the WE signal is removed.

3. To use the memory at an RTL design description, a user has to supply the memory with information about the address, data, control signals according to that specified in the memory timing specification.

4. In an algorithmic description, an array of variables (or an array of constants) is used to model linear structures such as an RAM (or a ROM) in hardware. The elements of a memory can be addressed using an index variable.

Memories

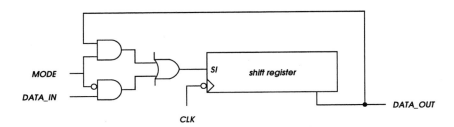

Figure X11.1

Exercises

1. Define a 32 × 8 ROM and a 32 × 8 RAM. What is the difference between the definitions of a ROM and a RAM?

2. In some digital circuit designs, multiple phase clocks are used. Suppose you are given a single phase clock, write a VHDL program to generate a four phase clock.

3. ROMs can be used to generate any waveform for a circuit. Repeat the previous design using a ROM to store the waveforms.

4. Figure X11.1 shows the block diagram for a recirculating shift register. Data enter the shift register from the serial input SI, which shifts into the most significant bit, the rest of the word will be shift right by one bit, and the least significant bit is recirculated back to the serial input through the control logic. This logic provides two modes of operations that are controlled by the *recirculation* input $MODE$. When $MODE$ equals 1, the data will be recirculated. In this mode, the $DATA\_IN$ is inhibited and has no effect on the register data. When $MODE$ equals 0, the $DATA\_IN$ signal is applied to the serial input. As clock pulses applied, new data will enter the register. Write a VHDL program for the circuit.

5. Figure X11.2 shows how seven recirculating shift registers are combined to store 7-bit ASCII-coded data for repetitive transmission to the circuits of video display. Suppose a video display system displays 24 lines of 80 characters. What is the storage capacity of the shift array memory.

6. Another way implementing the recirculating shift registers is using RAMs. Instead of shifting data along the registers, a pointer is used to indicate the location that a word is to be read or written into. Write a VHDL description for such a design.

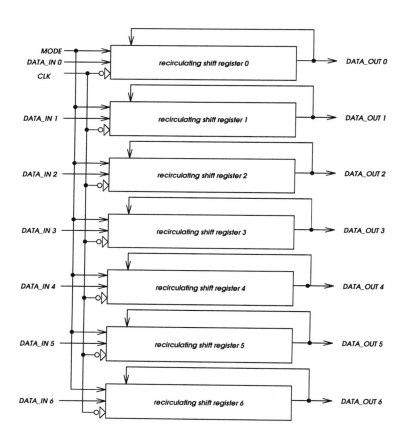

Figure X11.2

Memories 241

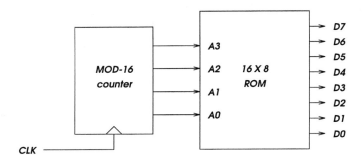

Figure X11.3

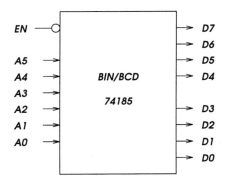

Figure X11.4

7. Figure X11.3 shows a 16 × 8 ROM with its address inputs driven by a MOD-16 counter so that the ROM addresses are incremented with each input pulse. Write a VHDL description for the circuit.

8. Figure X11.4 shows the logic symbol for the 74185 IC, which is an "off the shelf" ROM that is programmed as a *binary-to-BCD* converter. It converts a 6-bit binary input to a two-digit BCD output. Write a VHDL process to perform the function of 74185 without using ROM. Note that the output are from a set of tristate buffers which is controlled by an active low enable signal.

9. Repeat the previous problem using ROM.

10. Write a VHDL program which reads a 16 × 8 ROM, performs a bubble sort on the data of the ROM, and stores the results in a a 16 × 8 RAM.

11. Write a design description which implements 3x3 Tic-Tac-Toe (TTT) player.

12
VHDL SYNTHESIS

A lot of digital designs start at the behavior level, where designers describe the desired functionality algorithmically, where functionality is described by which operations must occur. During the design process, the system is partitioned into several less complex subsystems. Each subsystem is in turn partitioned into submodules until the submodules can either be mapped to existing modules (design re-use) or have well defined functionality and interfaces.

Modules with well-defined interfaces and functional behavior can be described using algorithms, as this is a natural way of describing a concept. An algorithmic description is easy to read, maintain, and modify. While the existing designs have well defined characteristics and structures, they can be described at the register transfer level or at the gate level.

Without the use of behavioral synthesis, an algorithmic specification must be manually translated to an RTL specification. This translation process is very time consuming and error prone. It was estimated that 40% of an IC's total design cycle is spent developing and verifying from conception to the RTL-level specification stage. Furthermore, because the RTL specification bears little resemblance to the algorithmic specification, any changes that are made to the algorithmic specification (and there are always changes), either for functionality or performance requirements, usually require a complete redesign. This iterative process greatly increases the design cycle.

To increase productivity at this stage of the design cycle, a synthesis tool should be able to take a design description at a mixed abstraction level (algorithmic, state machine, RT level, or gate level). Fig. 12.1 shows a design description containing a set of processes at different abstraction levels connected by signals.

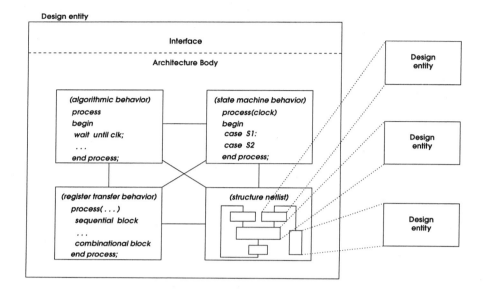

Figure 12.1 A mixed-level design description.

The target-system architecture is composed of interacting hardware processors in which communication is achieved using signal connections through some predefined protocols. At higher levels of abstraction, the processors are represented by processes, while at the gate level, the subsystems are represented as a set of concurrent assignments or component instantiations.

The methodology to be discussed in this chapter divides a design abstraction into four levels: algorithmic, finite state machine with datapath (FSMD), register transfer (RT), and gate levels. A synthesis step refers to a translation of a design description from one level of abstraction to a low level one. As the synthesis process goes, more and more details of a design appear. Fig. 12.2 shows the translation process from top to bottom. An algorithmic level description is converted into a FSMD level description by a scheduling algorithm; an FSMD level description is translated into a register transfer level description by an allocation algorithm; and a register transfer level description is converted into a gate level structure by an RTL synthesis tool.

VHDL Synthesis

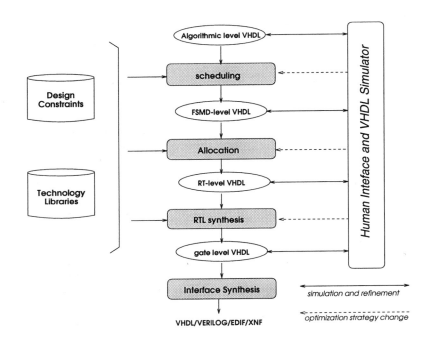

Figure 12.2 Synthesis from high level abstraction to low level.

In the following sections, we will discuss the major components in such a synthesis environment – VHDL design description, constraints, technology libraries, and synthesis tools.

12.1 VHDL DESIGN DESCRIPTIONS

A module can be described in one of the following four abstraction levels – algorithmic level, FSMD level, register transfer (RT) level, or gate level. All but gate level are behavior descriptions. A gate-level description consists of interconnected gate level cells.

Let's use a single pulser to illustrate the design description in the four different levels of abstraction. For the single pulser design, we have a debounced pushbutton, on(true) in the down position, off(false) in the up position. The problem is to devise a circuit to sense the depression of the button and assert an output signal for "exactly" one clock pulse. The system should not allow additional assertions of the output until after the operator has released the button. Let the input be DATA_IN and the output be DATA_OUT. Fig. 12.3 (a) shows the timing diagram for the circuit. The following shows the interface declaration of the single pulser.

```
entity PULSE is
port ( CLK              : in  BIT ;
       DATA_IN          : in BIT ;
       DATA_OUT         : out BIT ;
     ) ;
end ;
```

12.1.1 Algorithmic description

Algorithmic descriptions do not have states assigned, and need to be scheduled before they can be mapped to an FSMD description. Design behavior is expressed in a sequential language using sequential assignments for data transfer, control constructs for conditional execution, loop statements for iterative sequencing, and *wait* statements for signal inputs and outputs. This type of behavior comes closest to programs written in a standard imperative programming language.

Fig. 12.3 (b) shows an abstract algorithm of the single pulser circuit. The circuit checks on every leading edge of the clock to see if DATA_IN is '1'. If so, it sets the DATA_OUT to '1' and gets into a loop. In the loop, it sets the output to zero and checks again on every clock cycle to see if the button has been released (DATA_IN = '0'). If so, it exits the loop and returns to the beginning of the process.

```
architecture ALG of PULSE is
begin
P1 :   process
begin
   wait until CLOCK'EVENT and CLOCK = '1' and DATA_IN = '1' ;
   DATA_OUT <= '1' ;
   loop
     wait until CLOCK'EVENT and CLOCK = '1' ;
     DATA_OUT <= '0' ;
```

VHDL Synthesis

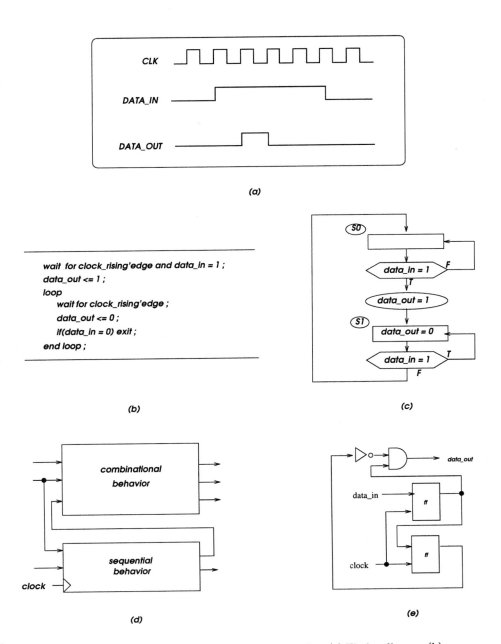

Figure 12.3 Design representation for a single pulser (a) Timing diagram (b) Algorithmic level (c) State machine level (d) Register transfer level (e) Gate level.

```
         exit when (DATA_IN = '0') ;
      end loop ;
   end process ;
   end ALG ;
```

12.1.2 FSMD description

To describe the single pulser circuit at FSMD level, we can partition it into two states as shown in Fig. 12.3 (c). In the first state, the circuit keeps comparing DATA_IN. Immediately after DATA_IN changes to '1', DATA_OUT will change to '1'. Then, the control of the circuit is transferred to state 1. At state 1, DATA_OUT is reset to '0' and the circuit stays in state 1 until DATA_IN changes to '0'. The following shows a state machine description in VHDL.

```
architecture FSMD of PULSE is
begin
process (CLK)
   type S_Type is (ST0 , ST1 ) ;
   variable STATE : S_Type := ST0 ;
begin
   if (CLK'EVENT and CLK = '1') then
      case STATE is
         when ST0 =>
            if (DATA_IN = '1') then
               DATA_OUT <= '1' ;
               STATE := ST1 ;
            else
               STATE := ST0 ;
            end if ;
         when ST1 =>
            DATA_OUT <= '0' ;
            if (not (DATA_IN = '1')) then
               STATE := ST0 ;
            else
               STATE := ST1 ;
            end if ;
```

```
      end case ;
    end if ;
  end process P1 ;
end FSMD ;
```

12.1.3 Register transfer description

The behavior of the output signals of an Register Transfer Level (RTL) circuit can be characterized to be synchronous and asynchronous as shown in Fig. 12.3 (d). A sequential behavior is one in which the output signals change values only on the clock edges; while a combinational behavior is one in which the output signals change values whenever inputs change value. In VHDL, sequential behaviors are modeled by a group of sequential statements enclosed by an **if** clock'edge statement; A combinational behavior is modeled by a group of statements that are evaluated sequentially whenever the signals in the sensitivity list change values.

For example, the following shows an RTL description of the single pulser circuit.

```
    architecture RTL of PULSE is
      signal TMP,A: bit ;
    begin
      process(CLOCK)
        begin
          if (CLOCK'EVENT and CLOCK = '1') then
            TMP <= DATA_IN ;
            A <= TMP ;
          end if ;
          DATA_OUT <= TMP and (not A ) ;
      end process ;
    end RTL ;
```

12.1.4 Gate level description

A gate-level design is composed of an interconnection of logic gates or macros. Fig. 12.3 (e) shows a gate level schematic. The following VHDL program is a gate level design of the single pulser. The gates are declared using component declaration statements and used by the component instantiation statements.

For example, the following shows a gate level description of the single pulser circuit.

```
architecture GATE of PULSE is
  signal NET1, NET2, NET3:  BIT ;
  component NOT_GATE port (I0:  in BIT; O: out BIT);
    end component ;
  component AND2_GATE port (I0, I1:  in BIT; O: out BIT);
    end component ;
  component DFF_GATE port (CK, D: in BIT; Q: out BIT);
    end component ;
begin
  U1:   NOT_GATE port map(I0=>net1, O=>net2) ;
  U2:   AND2_GATE port map(I0=>net2, I1=>net3, O=>data_out) ;
  U3:   DFF_GATE port map(CK=>CLOCK, D=>data_in, Q=>net3) ;
  U4:   DFF_GATE port map(CK=>CLOCK, D=>net3, Q=>net1) ;
END ;
```

12.2 CONSTRAINTS

Constraints are used to guide the optimization and mapping of a design towards feasible realizations in terms of area, performance, costs, testability, power consumption, and other physical limitations. They provide goals for the optimization and synthesis tools to achieve. Performance and area constraints are the two most common constraints.

For behavior synthesis, the area constraints are usually specified at the architectural level where a designer specifies the number of function units, registers and buses to be used on the RTL structure. The timing constraints are specified as the expected clock frequency of each clock signal.

12.3 TECHNOLOGY LIBRARY

Technology libraries contain all the information needed by the synthesis tool to make a correct choice in order to build the structure of a circuit. It contains the behavior of a cell and information such as the area of the cell, the timing

VHDL Synthesis

of the cell, the capacitance loading of the cell, etc. A technology library defines both rise and fall delay values for the basic cells. The example below shows a simple technology library description.

```
library (example)
  cell(AND2_gate)
    input I0, I1;
    output O;
    area := 2 ;
    function O := I0 & I1 ;
    capacitance I0,I1 := 1;
    pos_delay O : I0,I1 := 310 82 280 36;

  cell(NAND2_gate)
    input I0, I1;
    output O;
    area := 2 ;
    function O := !(I0 & I1) ;
    capacitance I0,I1 := 1;
    neg_delay O : I0,I1 := 90 41 70 29;

  cell(XOR_gate)
    input I0, I1;
    output O;
    area := 2 ;
    function O := I0 ^I1;
    capacitance I0,I1 := 2;
    non_unate_delay O : I0,I1 := 420 82 540 36;

  cell(DFR)
    input R,CK,D;
    output Q,QX;
    area := 4;
    function Q, !QX := posedge CK ? D : Q;
    setup D := 10 ;
    reset  R := active_low;
    capacitance  R,CK,D := 1;
```

```
pos_delay Q  : CK := 660 82 730 36;
pos_delay QX : CK := 880 82 870 36;
neg_delay Q  : R  := 580 36 0 0;
neg_delay QX : R  := 730 82 0 0;
```

The name of this technology library is called *example* and it contains four library cells – an *AND2_gate*, a *NAND2_gate*, an *XOR_gate*, and a *DFR_gate*. The first cell is named *AND2_gate*. It has two input pins $I0$ and $I1$, and one output pin O. The area of the cell is 2 units. The function of output pin O is an & (*and*) of $I0$ and $I1$. The loading capacitance for each input pin is 1 unit. The timing delay from $I0$ or $I1$ to O is based on a *pos_delay* calculation method with the following four parameters: D_r, R_r, D_f, R_f = (310, 82, 280, 36).

The delay calculation is based on the lumped RC model (also called the Elmore delay model). Each cell's timing is characterized by the four parameters: D_r, R_r, D_f, and R_f, where D_r represents the rising intrinsic delay, R_r represents the resistance for the signal to change from low to high, D_f represents the falling intrinsic delay, and R_f represents the resistance for the signal to discharge from high to low. Different gates use different delay calculation methods. In this example library, there are three delay calculation methods: *pos_delay*, *neg_delay*, and *non_unate_delay*.

- *pos_delay* means the rise delay at the output node is calculated by adding the input rise delay with the local rise delay, and the fall delay at the output node is calculated by adding the input fall delay with the local fall delay.

- *neg_delay* means the rise delay at the output node is calculated by adding the input rise delay with the local fall delay, and the fall delay at the output node is calculated by adding the input fall delay with the local rise delay.

- *non_unate_delay* means the rise (fall) delay at the output node is calculated by adding the input rise (fall) delay with the worst case of the local rise and fall delay.

For example, the timing characteristic for the AND2_gate is *pos_delay O : I0,I1 := 310 82 280 36;*. It means the rising (falling) delay of node O depends on inputs $I0$ and $I1$. We shall use *pos_delay* delay calculation. For output O to

VHDL Synthesis

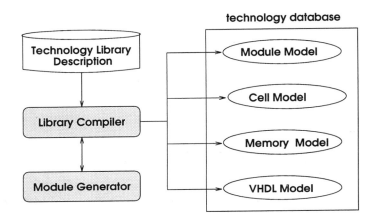

Figure 12.4 The library compiler.

rise from 0 to 1, the intrinsic delay is 310, and the resistance is 82, For output O to fall from 1 to 0, the intrinsic delay is 280, and the resistance is 36.

A technology library description is compiled into a technology database which will be used by synthesis tools or simulation tools. Since the primitive components for the behavior synthesis algorithm are RTL macros (adders, ALUs, registers, multiplexers, etc.), while a technology library may contain only basic cells (AND, OR, XOR, 1-bit multiplexer, etc.), we need to derive the timing and area information for these macros. The *library compiler* combined with a *module generator* are used to generate this information. Fig. 12.4 shows the function of the library compiler. For each technology library, the library compiler compiles the following information for the purpose of synthesis or simulation: cell model, module model, memory model, and VHDL model.

- The *cell model* contains the timing and area information of the basic cells described in the technology library specification. The information will be used by RTL synthesizer.

- The *module model* contains the timing and area information for the RTL macros. These RTL macros are designed on the fly by a module generator using the basic cells in the specified technology library.

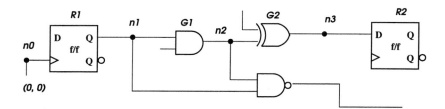

Figure 12.5 Delay calculation.

- The *memory model* contains information which is needed to estimate the timing for read and write access times of a memory. It is based on a set of parameters described in the header of the technology library.

- The *VHDL model* contains the VHDL description of the basic cell which is needed for simulation of gate level designs.

12.4 DELAY CALCULATION

In a CMOS technology, we model a component with its input capacitance and its output resistance. The delay of a combinational circuit is the propagation delay on the critical path through the circuit, that is, the longest delay from any input to any output.

Storage elements such as latches and flip-flops are used in a sequential circuit. To ensure proper operation, two timing constraints are imposed: the setup time (t_{setup}) and the hold time (t_{hold}). The setup and hold times specify time intervals before and after clock transitions for a data signal to be stable to ensure proper operation of the latch.

The clock cycle is determined by the worst case delay from a register to a register of the synthesized design. A typical register-to-register delay of a circuit is determined by the propagation delay of the source register, the wiring delay, the gate delay, and the set-up time of the destination register.

In the lumped RC model, the propagation delay of a node is calculated by lumping the resistances and capacitances of the node. Let us use the example shown in Fig. 12.5 to illustrate a delay calculation. Let ($t_r(n)$, $t_f(n)$) be the

VHDL Synthesis

worst case rising and falling delay of node n. First, we assume it is (0, 0) at node $n0$. The delay of path R1-G1-G2-R2 of Fig. 12.5 can be calculated as follows.

The first gate on the path is a DFR gate. The delay calculation method for a DFR gate is the *pos_delay* method. Hence

$$t_r(n1) = t_r(n0) + D_r(DFR) + R_r(DFR) \times C(n1)$$
$$= 0 + 660 + 82 \times 2 = 824, \text{ and}$$

$$t_f(n1) = 0 + 730 + 36 \times 2 = 802.$$

The second gate is an AND gate. The delay calculation method is also *pos_delay*. Therefore,
$$t_r(n2) = 824 + 310 + 82 \times 2 = 1298, \text{ and}$$

$$t_f(n2) = 802 + 280 + 36 \times 2 = 1154.$$

The delay calculation method for an XOR gate is *non_unate_delay* which means the delay is based on the worst case of the rise and fall input delay plus the intrinsic rise (or fall) delay. Hence, the delay at node $n3$ is

$$t_r(n3) = 1298 + 420 + 82 \times 1 = 1800, \text{ and}$$

$$t_f(n3) = 1298 + 540 + 36 \times 1 = 1874.$$

Finally, we reach register $R2$. Since there is a set up time requirement for the DFR gate, we should add the set up time to the delay of this path. So, the rise and fall delay for the path are

$$t_r(R1 \rightarrow G1 \rightarrow G2 \rightarrow R2) = 1800 + 10 = 1810, \text{ and}$$

$$t_f(R1 \rightarrow G1 \rightarrow G2 \rightarrow R2) = 1874 + 10 = 1884, \text{respectively.}$$

12.5 THE SYNTHESIS TOOL

Fig. 12.6 shows the major components of the MEBS system – a Library Compiler, a VHDL Compiler, a Scheduler, an Allocator, an RTL Synthesizer, an Area/Timing Estimator, a Netlist Generator, and a Design Flow Manager.

- The *Library Compiler* compiles a user defined technology library in text format into the system's technology database.

- The *VHDL compiler* compiles a VHDL description into the system's design library. It performs the following three subtasks:

    - The *VHDL Parser* analyzes a VHDL program, decomposes a VHDL file into library units, and stores the library units in the working library.

    - The *VHDL Model Checker* performs the synthesizability checking and determines the level of abstraction for each process in a VHDL program. Not all syntactically correct VHDL descriptions are synthesizable. For example, the following process is non-synthesizable.

    ```
    NS: process (a)
        begin
            c <= a and b ;
        end process;
    ```

    Intuitively, it can be realized by a two input AND gate. However, since signal c is not sensitive to signal b, the process does not behave like an AND gate. Thus, process NS is un-synthesizable. A correct specification is to put both signals a and b in the sensitivity list.

    - The *Pre-Synthesis Optimizer* performs a source level optimization including dead code elimination, constant propagation, common subexpression extraction, tree height reduction, and subprogram in-line expansion.

- The *Scheduler* converts an algorithmic-level description into an FSMD-level description. The scheduling algorithm performs code movement, code duplication, and operation chaining to minimize the number of states. An operation can be moved across several control flow blocks as long as it does not violate the semantics of the original program. An operation may be duplicated in different states to speed up the program execution. MEBS

VHDL Synthesis

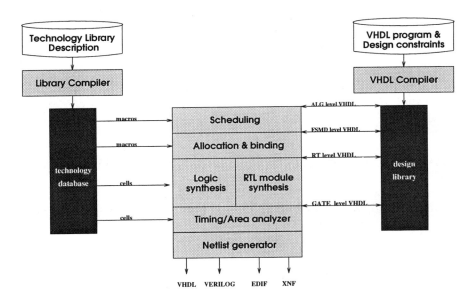

Figure 12.6 The MEBS system.

allows fixed-I/O scheduling where a designer specifies a cycle-by-cycle behavior of a circuit using **wait** statement, and block-based scheduling where a block of statements (without wait statements) will be cut into multiple states.

- The *Allocator* converts a FSMD-level description into an RT-level description. An RTL design may consist of the following components – data path, controller, and memory. Data path operations are extracted from a state description and a data path is synthesized. Arrays of variables(constants) are stored into RAM (ROM) and array references are converted into reading and writing memories. Controller controls the sequencing of the machine and issues commands to the data path and memories.

- The *RTL synthesizer* translates an RT-level description to netlist of gates of a specified technology library. A data path operation (such as addition, multiplexing, multiplication, etc.) can either be optimized by a module generator or by the logic synthesis tool specified as an option by a designer. The random logic is optimized using SIS – the logic synthesis tool developed at the University of California, Berkeley.

- The *Area/Timing Estimator* estimates the performance and the area of the design based on an Elmore time model using the area and delay information in the technology library.

- The *Design Flow Manager* is used to control the design flow of a design. It ensures the correctness of the dependencies of the design units in the design library.

12.6 DESIGN SPACE EXPLORATION

To illustrate how to use the MEBS system to explore a design space, we will use a differential equation solver as an example. The problem is to solve the following differential equation:

$$y'' + 3xy' + 3y = 0$$

where x and y are functions of variable u, i.e. $x = f(u)$, $y = g(u)$. The following iterative algorithm finds the values of x and y.

```
while (x<a) repeat
   xi := x  + dx ;
   ui := u - (3 * x * u * dx) - (3 * y * dx) ;
   yi := y + (u * dx) ;
   x := xi ; u := ui ; y := yi ;
end ;
```

The above algorithm can be translated into the following VHDL program. Here, Xi, Yi, Ui and DXi are inputs ports (8-bit numbers); Xo and Yo are output ports. When the READY signal changes to 1, the circuit starts the computation. After finishing the computation, it puts the results on the output ports and issues a DONE signal.

```
entity DIFFEQ is
   port(CLOCK, RESET : in  BIT;
        Xi, Yi, Ui, DXi   : in INTEGER range 0 to 255 ;
        Xo, Yo            : out INTEGER range 0 to 255 ;
        READY             : in BIT ;
        DONE              : out BIT := '0' ) ;
```

```
      end DIFFEQ ;
      architecture BE1 of DIFFEQ is
        signal x, y, dx, u :  integer range 0 to 255 ;
        constant A : integer range 0 to 255 := 1;
        -- mebs RESET RESET
      begin
      process
      begin
        wait until CLOCK'event and CLOCK = '1' and READY = '1';
        DONE <= '0';
        x <= Xi;
        y <= Yi;
        u <= Ui;
        dx <= DXi;
        while (x < A) loop
          wait until CLOCK = '1' ;
          x <= x + dx ;
          u <= u - (3 * x) * (u * dx) - (3 * y * dx) ;
          y <= y + (u * dx) ;
        end loop ;
        DONE <= '1' ;
        Xo <= x ;
        Yo <= y ;
      end process;
      end BE1;
```

The Pre-Synthesis Optimizer performs a source level optimization including dead code elimination, constant propagation, common subexpression extraction, tree height reduction, and subprogram in-line expansion. For example, The constant propagation algorithm will change the expression $(x < A)$ into $(x < 1)$. The size of a comparison-with-constant circuit is smaller than the size of a regular comparator. The common subexpression extraction algorithm may find the expression $(u * dx)$ can be extracted from the two signal assignment statements. Fig. 12.7 shows the control and data flow graph after the pre-synthesis optimization.

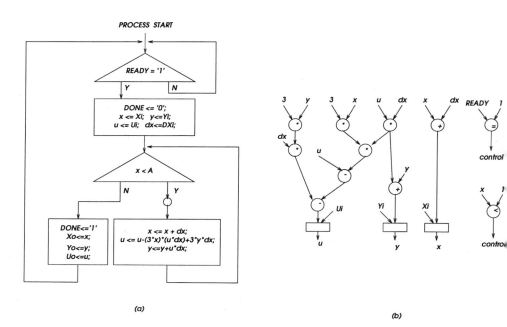

Figure 12.7 Differential equation solver (a) Control flow graph (b) Data flow graph.

VHDL Synthesis

constraints	synthesis options	# states	area	performance
1(ALU) 1(*) 1(<)	no mult-w-const	7	1453	154ns(6.49MHz)
1(ALU) 2(*) 1(<)	no mult-w-const	5	1594	127ns(7.87MHz)
1(ALU) 1(*) 1(<)	mult-w-const	5	1596	177ns(5.64MHz)
1(ALU) 2(*) 1(<)	mult-w-const	5	1917	189ns(5.29MHz)

Table 12.1   Design space exploration of the differential equation solver

Before scheduling, a designer can specify the technology library to build the circuit and set area/time constraints on the circuit. In addition, he can set the synthesis options to customize the optimization algorithms. For example, we can set *multiplier-with-constant* parameter to *share* or *no share* mode. A multiply-with-constant operation is an operation which multiplies a signal (or variable) with a constant. It can be done by a dedicated multiplier (which is smaller but non-sharable) and a generic multiplier (which is larger but sharable). The area of a dedicated multiplier is usually much smaller than a generic multiplier. For example, operation $(y*3)$ can be executed on a dedicated circuit which consists of just an adder.

Now, if we set the resource constraint to be one multiplier and one adder, and set the *multiplier-with-constant* parameter to the *share* mode. We will obtain an FSMD with 7 states. On the other hand, if we set the resource constraint to two multiplier and one ALU, the resulting FSMD description will have only 5 control states.

If we set the *multiplier-with-constant* parameter to *noshare* mode, we will obtain another set of results. Table 12.1 shows the results for the four different design decisions. In this example, using a dedicate multiplier does not give a better result than a generic multiplier. Fig. 12.8 shows the design space curve of the circuit. Note that we can trim two solutions from the design space because they are apparently inferior to the other two solutions.

Another example to show design space tradeoff is a greatest common divisor (GCD) calculator. The following shows the VHDL description at algorithmic level. The circuit starts computing the GCD of Xi and Yi when READY signal becomes high. When the computation finishs, it puts the result on the output port and issues a DONE signal.

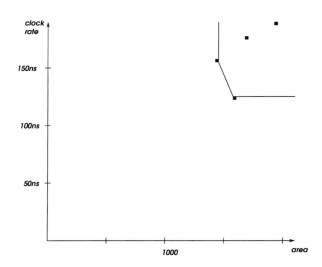

Figure 12.8 Design space curve of the differential equation solver

```
entity GCD is
  port (CLOCK : in  BIT ;
        RESET : in  BIT ;
        Xi,Yi :  in   INTEGER range 0 to 255 ;
        READY : in  BIT ;
        DONE  : out BIT := '0' ;
        OU    : out INTEGER range 0 to 255 := 0) ;
end GCD ;
architecture BHV of GCD is
-- mebs reset_high reset
begin
  process
    variable X, Y : INTEGER range 0 to 255 ;
  begin
    wait until (CLOCK'event and CLOCK = '1' and READY = '1')
;
    DONE <= '0' ;
    X := Xi ;
```

VHDL Synthesis

constraints	# states	multiplexors	area	performance
1(−)1(<)	2	4	485	149ns(6.7MHz)
2(−)1(<)	2	2	508	76ns(13.5MHz)

Table 12.2 Design tradeoff of the GCD calculator

```
      Y := Yi ;
      while (X /= Y) loop
         wait until (CLOCK'event and CLOCK = '1') ;
         if (X < Y) then
            Y := Y - X ;
         else
            X := X - Y ;
         end if ;
      end loop ;
      DONE <= '1' ;
      OU <= X ;
   end process ;
end BHV ;
```

Fig. 12.9 shows the control/data flow graph of the algorithmic description. There are two subtraction operations in the program. One is in *then* clause and the other is in the *else* clause. The scheduling produces the same result with 2 control states for the two constraints in Table 12.9. During the allocation process, the allocator will assign both subtraction operations to a single subtractor for the one with one subtractor constraint. Two multiplexors will be added to the inputs of the subtractor because of the sharing of the two subtractions. On the other hand, the two subtraction operations will be assigned to two different subtractors if the constraint is two subtractors. Because no multiplexors are needed in front of the subtractor, the critical path is much shorter than the other one. Although there is an extra subtractor for the second case, its area does not increase much because it uses less multiplexors.

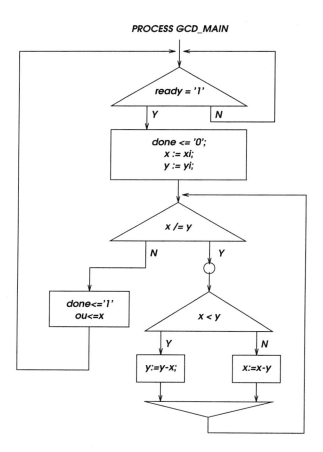

Figure 12.9 Control/Data flow graph of the GCD calculator

VHDL Synthesis

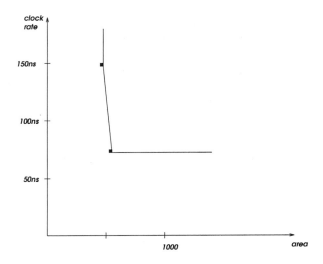

Figure 12.10 Design space curve of the GCD calculator

12.7 SYNTHESIS DIRECTIVES

MEBS provides several system defined *directives* that can be used to direct the translation process, specify the asynchronous set/reset signal, provide a means of mapping subprograms to hardware components and more.

Synthesis comments begin just as regular comments, with two hyphens (– –). If the word following these characters is **mebs**, the remaining comment text is interpreted as a directive.

12.7.1 Synthesis Off and On Directive

These directives stop and start the synthesis of the VHDL source file. The syntax is:

    – – **mebs synthesis_off**
    – – **mebs synthesis_on**

In VHDL, some statements are non-synthesizable, such as FILE declaration and file-access subroutines in package TEXTIO. These statements, however, are very useful for simulation and debugging but are not accepted by the MEBS compiler. We can use this directive to mask out those segments which we don't want the synthesizer to see but want the simulator to process. Translation is enabled at the beginning of each VHDL source file. *Synthesis_off* and *synthesis_on* can be used anywhere in the text.

The following example shows a design description for a multiplexer and its test bench. Both processes are described within the same architecture. Since the testbench is not needed for synthesis, we include it in the VHDL program but set a *synthesis off* directive to disable it from being synthesized.

```
library IEEE ;
use IEEE.NUMERIC_BIT.all ;
entity MUX is
  port (X0i :   buffer UNSIGNED(3 downto 0) ;
  X1i :   buffer UNSIGNED(3 downto 0) ;
  Z2i :   buffer UNSIGNED(3 downto 0) ;
  SEL : buffer BIT ) ;
end MUX;
architecture BHV of MUX is
begin
    MUX2 :
    process(SEL,X0i,X1i)
    begin
       case SEL   is
          when '0' => Z2i <= X0i ;
          when '1' => Z2i <= X1i ;
       end case ;
    end process ;
    -- mebs synthesis_off
    -- This section will be ignored by synthesizer
    X0i <= "0011" ;
    X1i <= "0101" ;
    process
    begin
```

VHDL Synthesis

```
        SEL <= '0' ;
        wait for 10 ns ;
        SEL <= '1' ;
        wait for 10 ns ;
        assert (FALSE)
            report "MUXs : Testing Complete !!!    "
            severity failure ;
    end process ;
    -- mebs synthesis_on
end BHV ;
```

MEBS supports another directives, *ignore_begin* and *ignore_end* to control the synthesizer on or off. The syntax is:

- − − **mebs ignore_begin**
- − − **mebs ignore_end**

The difference between these two sets of directives is that the source code will be kept after synthesis when they are masked by *synthesis_off* and *synthesis_on*. The code will be ignored by the VHDL parser if they are masked by *ignore_begin* and *ignore_end*. Using directives *ignore_begin* and *ignore_end* is similar to use /* and */ in C language. You may use *ignore_begin* and *ignore_end* anywhere in your VHDL source code. However, *synthesis_off* and *synthesis_on* can only be used in the declaration_part, concurrent_statement_part, or the sequential_statement_part.

12.7.2 Asynchronous Set/Reset Directives

These directives specify the asynchronous set/reset signal of the circuit module to be synthesized. They are only useful for processes at the algorithmic level. The syntax is:

- − − **mebs reset_high** *reset_signal_name*
- − − **mebs reset_low** *reset_signal_name*
- − − **mebs reset** *reset_signal_name*

The asynchronous reset signal resets or sets initial values to certain signals after the circuit starts or resets. Only one asynchronous reset signal is allowed in a process. The data type of the asynchronous reset must be BIT or STD_ULOGIC data type. It can only be specified in the declaration part. The **mebs reset_high** (**mebs reset_low**) directive means the signal is active high (low). Default (**mebs reset**) is active high.

With this directive, we can tell the synthesizer to set or reset certain registers after the asynchronous reset signal is set to high or low.

The following VHDL description shows how to use the asynchronous reset directive to specify an asynchronous reset signal with a name RESET at the algorithmic level. When RESET becomes high, the signals which have an initial value will be set to that value. In this case, the output signal CountOut will be set to 0.

```
entity Counter is
   port (CLOCK    : in     BIT ;
         RESET    : in     BIT ;
         CountOut : buffer INTEGER range 0 to 15 := 0 ) ;
end Counter;
architecture Algorithm of Counter is
   -- mebs reset_high reset
begin
   process
   begin
      wait until CLOCK'event and CLOCK = '1' ;
      CountOut <= (CountOut + 1) mod 16 ;
   end process;
end Algorithm ;
```

For a process at the level of FSMD, RTL, or Gate level, we no longer need this directive. The synthesis system can automatically detect the asynchronous reset signal. In the following example, RESET signal will be detected by the synthesis tool as the asynchronous signal.

```
architecture FSMD of Counter is
begin
   process (RESET, CLOCK)
```

```
    begin
       if (RESET = '1') then
          CountOut <= 0;
       elsif (CLOCK'event and CLOCK = '1') then
          CountOut <= (CountOut + 1) mod 16 ;
       end if;
    end process ;
 end FSMD ;
```

Synchronous reset signals are treated as regular signals which do not have to be specified.

12.7.3 Function Directives

These directives are used to specify the synthesis and mapping methods of a subprogram. The syntax is:

```
-- mebs inline
-- mebs module
-- mebs tri_state
-- mebs wire_or
-- mebs wire_and
```

The first directive (**inline**) specifies that the subprogram is to be synthesized by an in-line expansion method (a default method) where the subprogram is expanded into the main program. The second directive (**module**) specifies that the subprogram is to be synthesized independently as a module and will be treated as a function unit. A designer can describe a special and complicated function unit by this method. Only a function can be synthesized by the second method and it must be a combinational circuit.

The following example defines a module function for an 8-bit ALU.

```
-- mebs module
function ALU (I,J:BIT8; CT:BIT) return BIT8 is
   variable O : BIT8 ;
begin
```

```
        if (CT = '0') then
           O := I + J ;
        else
           O := I - J ;
        end if ;
        return O ;
    end ALU ;
```

The other three directives are used to specify the synthesis method of a resolution function. MEBS uses the three-state buffer to implement the **tri_state** resolution function, and uses or-gate, and and-gate to implement the **wire_or**, and **wire_and** resolution functions.

12.7.4 State Variable Directives

This directive specifies the name of the state variable of a VHDL process at the FSMD level. Without this directive, the synthesizer will still be able to synthesize a functionally equivalent circuit. However, the synthesizer can do more optimization to reduce the size of the circuit with this directive. The syntax is:

  $-$ $-$ **mebs state_var** *state_variable_name* { , *state_variable_name* }

Similar to the directive of asynchronous reset signal, this directive must be specified in the declaration part of each process. This directive is used to tell the synthesizer that the following variables are state variables in state machine based descriptions. The following example describes a state machine with a state variable named "State".

```
    process (RESET, CLK)
       type S_Type is (ST0, ST1) ;
       variable State :  S_Type := ST0 ;
       -- mebs state_var State
    begin
       if (RESET = '1') then
          Z <= '0' ;
          STATE := ST0 ;
```

```
            elsif (CLK'event and CLK = '1') then
               case STATE is
                  when ST0 =>
                     if (X = '1') then
                        Z <= '1';
                        State := ST1 ;
                     else
                        State := ST0 ;
                     end if;
                  when ST1 =>
                     if (X = '1') then
                        Z <= '0';
                        State := ST0 ;
                     else
                        State := ST1 ;
                     end if;
               end case;
            end if;
         end process;
```

The following example shows a VHDL program which specifies two state variables in a process.

```
      process (CLOCK, RESET, START, MEBS_tmp1, MEBS_tmp0)
         type S_Type is (ST0, ST1) ;
         variable State,Next_State :  S_Type   := ST0 ;
         -- mebs state_var State, Next_State
      begin
         if (RESET = '1') then
            State := ST0 ;
         elsif (CLOCK'event and CLOCK = '1') then
            State := Next_State ;
         end if;
         -- combinational circuit part
         case State is
```

```
            when ST0 =>
               if (START = '1') then
                  S0 <= SEL1 ;
                  Next_State := ST1 ;
               else
                  Next_State := ST0 ;
               end if ;
            when ST1 =>
               if (MEBS_tmp1 and (not MEBS_tmp0) then
                  ENO <= TRUE ;
                  Next_State := ST1 ;
               elsif (MEBS_tmp1 and MEBS_tmp0) then
                  S2 <= SEL1 ;
                  Next_State := ST1 ;
               else
                  Next_State := ST0 ;
               end if ;
         end case ;
      end process ;
```

12.7.5 Don't Care Value Directives

This directive specifies that the value of a signal in the list will only be valid in the clock cycle it is assigned. We don't care the value of the signals at other clock cycles.

$$- - \textbf{mebs dont\_care} \; \textit{don't\_care\_signals} \; \{ \, , \, \textit{don't\_care\_signals} \, \}$$

In VHDL, a signal will keep its current value until the next time it is updated. In a description of a sequential circuit, however, a signal may be valid when it is specified. If it is not specified, it is treated as don't care. We use this directive to specify these signals. This information can be translated into don't cares in the logic equations to assist the sequential and/or RTL synthesis tool to obtain a smaller circuit. In a synthesized circuit, these signals are usually the control signals of multiplexers or enable signals of registers. The directive may result in synthesis result and simulation result mismatch.

VHDL Synthesis

In the following VHDL program, the value of signal Z is don't care when the current state is ST1 and the condition "(Ctrl = '1')" is true. Similar interpretations can be applied to all output signals.

```
process (CLOCK, RESET, Ctrl)
    type S_Type is (ST0, ST1) ;
    variable State, Next_State :   S_Type ;
    -- mebs state_var State, Next_State
begin
    if (RESET = '1') then
        State := ST0 ;
    elsif (CLOCK'event and CLOCK = '1') then
        State := Next_State ;
    end if ;
    -- mebs dont_care Z
    case State is
        when ST0 =>
            if (Ctrl = '1') then
                Z <= '1' ;
                Next_State := ST1 ;
            else
                Next_State := ST0 ;
            end if ;
        when ST1 =>
            if (Ctrl = '1') then
                Next_State := ST1 ;
                -- Z gets a don't care value here
            else
                Z <= '0' ;
                Next_State := ST0 ;
            end if ;
    end case ;
end process ;
```

12.7.6 Register Array Directives

This directive specifies that the following array type is interpreted as a register array not as a RAM or ROM.

    -- mebs register_array

The unconstrainted array declaration after this directive will be represented as a register declaration. The following example shows how to use the register array directive.

```
type MEM_Array is array (NATURAL range <>) of BIT ;
-- mebs register_array
type REG_Array is array (NATURAL range <>) of BIT ;
variable A : MEM_Array(7 downto 0) ;
variable B : REG_Array(7 downto 0) ;
```

The variable A will be implemented by a 8×1 RAM module, It can be accessed one element at a time since a unique address for each element is required. The variable B will be implemented by a 8-bit register, so the slice access (multiple bits or whole array) is allowed. The predefined type BIT_VECTOR and IEEE standard types STD_ULOGIC_VECTOR, STD_LOGIC_VECTOR, UNSIGNED, SIGNED are interpreted as register array types in the MEBS system.

12.8 SUMMARY

1. Algorithmic descriptions do not have states assigned, and need to be scheduled before they can be mapped to an FSMD description. Design behavior is expressed in a sequential language using sequential assignments for data transfer, control constructs for conditional execution, loop statements for iterative sequencing, and *wait* statements for signal inputs and outputs.

2. An FSMD is an extension of a traditional FSM with an addition of a set of variables and a set of operations on the variables. Within a state

expression, we can perform comparison, arithmetic, or logic operations on these variables.

3. A Register Transfer Level (RTL) description is characterized by sequential behavior and combinational behavior.

4. Synthesis is a process of transforming a design described in one level of abstraction down to a lower level one.

5. A technology library contains a set of cells with information such as the area of the cell, the timing of the cell, the capacitance loading of the cell, etc.

6. A timing analyzer calculates the delay between two points based on the timing information provided in a technology library.

Exercises

1. Write an algorithmic design description for an 8-bit self-correcting ring counter whose states are 11111110, 11111101, ..., 01111111.

2. Manually design the 8-bit self-correcting ring counter in the previous problem down to a design at the register transfer level, and describe it in VHDL.

3. Translate the FSM and the FSMD of the modulo-3 divider in Table 9.1 into corresponding VHDL descriptions.

4. Manually convert the FSM and the FSMD of the module-3 divider in Table 9.1 into gate level. Draw a gate level schematic for each one.

5. Manually synthesize the following program segment into hardware.
   ```
   if X ≠ Y then
       Z <= A ;
   else
       Z <= '0' ;
   end if ;
   ```

6. Manually synthesize the following program segment into hardware.
   ```
   if CLK'event and CLK ='1' then
       Z <= '1';
   elsif X = Y then
       Z <= '0' ;
   end if ;
   ```

7. Given the following VHDL program. Assume X, Y, A, B, C and D are all 8-bit numbers. What is the synthesis result if the resource constraints are an 8-bit comparator and an 8-bit ALU? What if the resource constraints are an 8-bit comparator, an adder and a subtrator.
   ```
   if (X > Y) then
       Z <= A + B ;
   else
       Z <= C - D;
   end
   ```

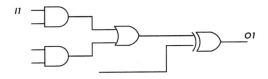

Figure X12.1

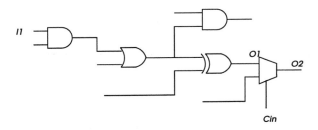

Figure X12.2

8. Given the circuit schematic in Figure X12.1, calculate the delay from input I1 to O1 using the delay information provided by the PRIMITIVE technology library in Appendix B.3. Assume the initial rising and falling delay of the input $I1$ is $(0, 0)$.

9. Given the circuit schematic in Figure X12.2, calculate the delay from input I1 to O2 using the delay information provided by the PRIMITIVE technology library in Appendix B.3. Assume the initial rising and falling delay of the input $I1$ is $(0, 0)$.

10. Given the circuit schematic in Figure X12.3, calculate the delay from register R1 and register R2 using the delay information provided by the PRIMITIVE technology library in Appendix B.3. Assume the initial rising and falling delay of the CLK is $(0, 0)$.

11. Given the circuit schematic in Figure X12.4, calculate the delay from the input to the output using the delay information provided by the PRIMITIVE technology library in Appendix B.3. Assume the initial rising and falling of the CLK is $(0, 0)$.

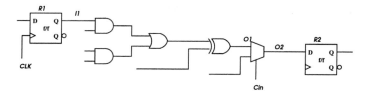

Figure X12.3

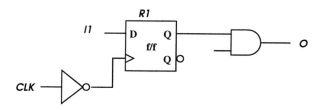

Figure X12.4

13
WRITING EFFICIENT VHDL DESCRIPTIONS

There are several ways of describing a design in VHDL which are functionally equivalent, yet are mapped to circuits of different sizes. Although certain transformation rules can be applied to optimize some "redundant descriptions", it is generally a designer's responsibility to develop a design description which will lead to an efficient circuit implementation. Take the following VHDL program as an example. TWO is declared as a signal in Process_A and a constant 2 is assigned to it. In Process_B, it is multiplied by signal B and the result is stored in signal C. Since TWO is assigned outside the process Mult_By_Two, the synthesizer may not able to recognize that TWO can be reduced to a constant. Thus, a regular multiplier will be synthesized to the assignment statement in Process_B. Fig. 13.1(a) shows a circuit using a regular multiplier for the operation.

```
    signal TWO, B, C : INTEGER range 0 to 15 ;
    . . .
    Process_A: process
    begin
        . . .
        TWO <= 2 ;
        . . .
        wait on CLK='1';
    end process ;
    . . .
    Process_B: process( CLK )
    begin
        if (CLK'event and CLK = '1') then
```

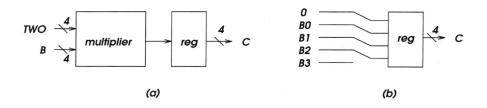

Figure 13.1 Schematics for multiply with constants.

```
      C <= B * TWO ;
   end if ;
end process ;
```

On the other hand, if we know TWO is a constant, we can declare it as a constant. The synthesizer will use an efficient circuit to implement a multiplication of a signal with a constant. For example, the implementation of a multiplication with a constant 2 can be zero cost by shifting the input, by one bit as shown in Fig. 13.1(b).

```
constant TWO:   INTEGER range 0 to 15 := 2 ;
signal B, C : INTEGER range 0 to 15 ;
Process_B: process ( CLK )
begin
   if (CLK'event and CLK = '1') then
      C <= B * TWO ;
   end if ;
end process ;
```

In the following sections, we enumerate some guidelines to write an efficient design description.

13.1 SOFTWARE TO HARDWARE MAPPING

During the synthesis process, each concurrent statement of a VHDL program such as a concurrent signal assignment, a process statement, or a component

Writing Efficient VHDL Descriptions 281

instantiation statement, is synthesized as a piece of logic. The signals that communicate with the concurrent statements are synthesized as wires. Inside each process, a signal may need a register to hold the value if it is assigned inside the process. A variable may correspond to a wire or a register depending on its lifetime. A statement or an operator in a process may be mapped to hardware such as adders, subtractors, multiplexers, decoders, latches, etc.

A designer should have some concept on how a software statement may be mapped to a hardware logic. Let's use the following VHDL description to illustrate the hardware mapping concept.

```
signal S1 :  bit ;
signal S2, S3, OU : integer range 0 to 255 ;
process
   variable TEMP1, TEMP2 :  integer range 0 to 63 ;
   variable A, B, C: integer range 0 to 255 ;
begin
   wait until (CLK'event and CLK='1' and S1 = '1') ;
   if (TEMP1  < 8 ) then
     S2 <= A + B ;
   else
     S2 <= C ;
   end if ;
   if (TEMP2 > TEMP1) then
     OU <= S3 + A ;
   end ;
end process ;
```

In this program, each statement represents a certain meaning to the synthesized circuit. For the VHDL program, we know we will need

- a comparator that compares the value of TEMP1 and 8,
- an adder that adds A and B,
- a register which holds the value of S2,
- a multiplexer that controls the final value of S2,
- an adder that adds S3 and A,

- a comparator that compares the value of TEMP1 and TEMP2, and
- a register that holds the value of OU.

A synthesizer will use one or two adders to execute the two addition operations depending on the resource constraints. If one adder is used, the two addition operations shall not be executed at the same clock cycle. Since the inputs to the single adder may come from different sources, multiplexers may be introduced to determine the right input signals to the input ports of the adder. If variable A is alive in more than one clock cycle, it has to be stored in a register, even though it is used only inside a process. A circuit for comparing a variable (signal) and a constant is different from a circuit for comparing two variables (or signals). It is better for a designer to have some knowledge on hardware mapping.

13.2 VARIABLES AND SIGNALS

Variables are defined inside a process and are used to store intermediate results in VHDL. The value of a variable has to be stored in a register if its lifetime spans more than one clock cycle A register can be used to store multiple variables if their lifetime spans do not overlap.

Signals have to maintain their values unless they are assigned with new values in VHDL. Signal assignments within an "if signal'edge" clause or in an algorithmic process must use flip-flops to hold the values of the target signals. Therefore each signal needs a dedicated register in a process.

Avoid using signals to store intermediate results. For example, the following code segment generate a three flip-flops:

```
signal Data_In, Data_Out, CLK : BIT ;
process
   variable A, B : BIT ;
   if (CLK'event and CLK='1') then
     Data_Out <= A ;
     A := B ;
     B := Data_In ;
   end if ;
```

Writing Efficient VHDL Descriptions

```
    end process ;
```

In this case, the variables A and B are both used before they are assigned. Since a process is itself an infinitive loop. They have to pass their values assigned in last cycle to this cycle. If the variables are assigned before they are used, the circuit will be different. For example,

```
    signal Data_In, Data_Out, CLK : BIT ;
    process
        variable A, B : BIT ;
        if (CLK'event and CLK='1') then
            B := Data_In ;
            A := B ;
            Data_Out <= A ;
        end if ;
    end process ;
```

In this case, A and B are assigned before used, and therefore do not generate flip-flop. Instead, each of them is simply a wire. Only one flip-flop will be generated because of the signal assignment in the process.

13.3 USING MINIMUM BIT WIDTH

If we know exactly what the range of a signal or variable is, we should declare this explicitly since it reduces the hardware cost and increases the performance. A variable (or signal) which is declared as an integer without specifying the range may be synthesized using a default value of 32. The impact of the declaration will not only affect the size of registers, it will also affect the size of the bitwidth of the function units. For example, the interface of the following VHDL design shows the ports A, B and OU are 8-bit integers, and C is a 4-bit integer. In the architecture design EX1, TEMP is a temporary storage location for either B or C depending on the result of the comparison. Since TEMP is declared as an 8-bit integer, we will need an 8-bit multiplexer for the **if** statement and an 8-bit full adder for the addition statement. On the other hand, in the architecture design EX2, TEMP is declared as a 4-bit variable. We need only a 4-bit multiplexer and the upper four bit of the adder can be replaced by simpler logic. Therefore, the synthesized circuit will be smaller.

```
    entity minimum_bitwidth is
```

```vhdl
    port (A, B : in   integer range 0 to 255 ;
          C    : in integer range 0 to 15 ;
          OU   : out integer range 0 to 255 := 0) ;
end minimum_bitwidth;
architecture EX1 of minimum_bitwidth is
begin
  process(A, B, C)
    variable TEMP : integer range 0 to 255 ;
  begin
    if (B < 8 ) then
      TEMP := B ;
    else
      TEMP := C ;
    end if ;
    OU <= A + TEMP ;
  end process ;
end EX1 ;
architecture EX2 of minimum_bitwidth is
begin
 efficient:
  process(A, B, C)
    variable TEMP : integer range 0 to 15 ;
  begin
    if (B < 8 ) then
      TEMP := B ;
    else
      TEMP := C ;
    end if ;
    OU <= A + TEMP ;
  end process ;
end EX2 ;
```

13.4 USING EFFECTIVE ALGORITHMS

In some circuit designs, a circuit has to wait for a constant delay before it proceeds to the next event. For example, in a traffic light controller, we need a long timer and a short timer to control the red light and the yellow light, respectively. One way to implement it is using counters. We can use the conventional binary counters or the Linear-Feedback-Shift-Register to implement the delay circuit.

Conventional binary counters use complex or wide fan-in logic to generate high end carry signals. The delay circuit works in the following way. First, the counter is loaded with an initial value. Then, the counter is incremented by one at every clock cycle. When the count reaches a certain value we generate a time out signal. This scheme preserves the entire counting sequence. However, it has several drawbacks. The first drawback is a large area overhead. It needs a binary counter (an adder) and a register to implement the timer as shown in Fig. 13.2 (a)). The second drawback is a long delay of a binary counter. It cannot be used in a high performance design (with very small clock delay) due to the critical path through the counter.

Another solution is to use a Linear Feedback Shift Register (LFSR). An n-bit LFSR counter can have a maximum sequence length of $2^n - 1$. It goes through all possible code permutations except one, which is a lock-up state. LFSR uses a much simpler structure: a n-bit shift register and an XNOR (or an XOR) gate in the feedback path from the last output to the first input as shown in Fig. 13.2 (b). The XNOR makes the lock-up state: the all-ones state (An XOR gate would make it the all zeros state). The LFSR scheme sacrifices the binary counting sequence, but it has very good performance within several nano seconds per clock cycle.

The following example shows the two different implementations of a constant delay circuit.

```
Delay_counter:
process (reset_delay,clk)
begin
    if (reset_delay = '1') then
        count := "0000" ;
    elsif (clk'event and clk = '1') then
        count := count + 1 ;
    end if;
```

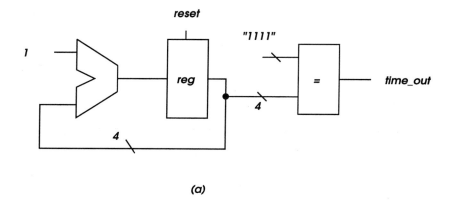

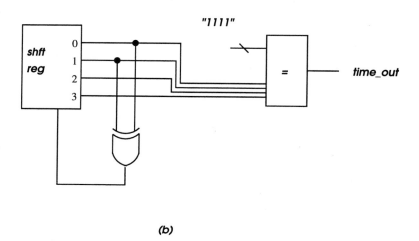

Figure 13.2 Timer implementation using (a) a conventional binary counter (b) an Linear Feedback Shift Register.

```
            time_out_counter <= count = "1111";
    end process;
    Delay_LFSR:
    process (reset_delay,clk)
        variable count :  unsigned(3 downto 0);
    begin
        if (reset_delay = '1') then
            count := "0001";
        elsif (clk'event and clk = '1') then
            count := (count(1) xor count(0)) & count(3 downto 1);
        end if;
        time_out_lfsr <=  count = "0011";
    end process;
```

13.5 SHARING COMPLEX OPERATORS USING MODULE FUNCTIONS

A subprogram can be used to define a complex functional unit which takes more than two inputs and produces more than one output. If such a function unit will be used in several locations in a program, we can define it as a module function. This allows an expensive operator to be shared in the circuit. By specifying a function subprogram as a module function, the synthesizer will synthesize it independently and treat it as a function unit module such as an adder or a multiplier. Two function calls can then share the same function unit if they are mutually exclusive. The following example shows a definition of a BCD incrementer.

```
    subtype bcd4_type is unsigned(3 downto 0);
    subtype bcd5_type is unsigned(4 downto 0);
    constant BCD5_1      :  bcd5_type := B"0_0001";
    constant BCD5_7      :  bcd5_type := B"0_0111";
    constant BCD5_9      :  bcd5_type := B"0_1001";
    -- mebs module
    function BCD_INC (L : in BCD4_TYPE) return BCD5_TYPE is
        variable V : BCD5_TYPE ;
    begin
```

```
            V := L + bcd5_1 ;
         if (V > bcd5_9) then
               V := V + 6;
      end if;
      return (V) ;
   end BCD_INC ;
```

13.6 SPECIFYING DON'T CARE CONDITIONS

Using don't care conditions allows the optimizer to fully utilize don't care property to minimize a circuit. One can greatly reduce circuit size by using *don't cares*. The following example shows two processes for the seven segment display of a BCD number. The one with don't care specification will be synthesized into a smaller circuit.

```
library ieee ;
use ieee.std_logic_1164.all ;
. . .
signal A: STD_LOGIC_VECTOR(3 downto 0) ;
signal OU: STD_LOGIC_VECTOR(6 downto 0)) ;
. . .
DECODE_BCD1:  process (A)
   begin
      case (A) is
         when "0000" => OU <= "1000000" ;   -- 0
         when "0001" => OU <= "1111001" ;   -- 1
         when "0010" => OU <= "0100100" ;   -- 2
         when "0011" => OU <= "0110000" ;   -- 3
         when "0100" => OU <= "0011001" ;   -- 4
         when "0101" => OU <= "0010010" ;   -- 5
         when "0110" => OU <= "0000010" ;   -- 6
         when "0111" => OU <= "1111000" ;   -- 7
         when "1000" => OU <= "0000000" ;   -- 8
         when "1001" => OU <= "0010000" ;   -- 9
         when others => OU <= "0000000" ;
```

```
          end case ;
      end process DECODE_BCD1 ;
   DECODE_BCD2:  process (A)
      begin
         case (A) is
            when "0000" => OU <= "1000000";   -- 0
            when "0001" => OU <= "1111001";   -- 1
            when "0010" => OU <= "0100100";   -- 2
            when "0011" => OU <= "0110000";   -- 3
            when "0100" => OU <= "0011001";   -- 4
            when "0101" => OU <= "0010010";   -- 5
            when "0110" => OU <= "0000010";   -- 6
            when "0111" => OU <= "1111000";   -- 7
            when "1000" => OU <= "0000000";   -- 8
            when "1001" => OU <= "0010000";   -- 9
            when others => OU <= "XXXXXXX";
         end case ;
      end process DECODE_BCD2 ;
```

13.7 WRITING LOW LEVEL CODE

The lower the level of abstraction of a VHDL description, the less effort for the synthesizer to generate a result. If the best circuit structure is already known, a user may describe it at a lower level of abstraction. In this way, a designer can guarantee to achieve the desired circuit. For example, both the following two functions implement a BCD incrementer. To synthesize the first function, we need two adders (of different sizes) and a comparator; while for the second function, we use only one adder and no comparator is needed.

```
      subtype bcd4_type is unsigned(3 downto 0);
      subtype bcd5_type is unsigned(4 downto 0);
      constant BCD5_1    :   bcd5_type := B"0_0001";
      constant BCD5_7    :   bcd5_type := B"0_0111";
      constant BCD5_9    :   bcd5_type := B"0_1001";
      -- mebs module
      function BCD_INC_EFFICIENT (L : in BCD4_TYPE) return BCD5_TYPE is
```

```vhdl
        variable V : BCD5_TYPE ;
    begin
        V := L + bcd5_1 ;
        if (V > bcd5_9) then
            V := V + 6;
        end if;
        return (V) ;
    end BCD_INC_EFFICIENT ;
    -- mebs module
    function BCD_INC_INEFFICIENT (L : in BCD4_TYPE) return BCD5_TYPE
is
        variable V, V1, V2 :  BCD5_TYPE ;
    begin
        V1 := L + bcd5_1 ;
        V2 := L + bcd5_7 ;
        case V2(4) IS
            when '0' => V := V1 ;
            when '1' => V := V2 ;
        end case ;
        return (V) ;
    end BCD_INC_INEFFICIENT ;
```

13.8 SUMMARY

1. There are several ways of describing a design in VHDL which are functionally equivalent, yet are mapped to circuits of different sizes.

2. A VHDL program which considers hardware mapping can achieve better results than those don't.

3. If we know exactly what the range of a signal or variable is, we should declare this explicitly.

4. Module functions allow one to define a complex functional unit.

5. Using don't care conditions allows the optimizer to further reduce the circuit size.

6. The lower the level of abstraction of a VHDL description, the less effort for the synthesizer to generate a result. If the best circuit structure is already known, a user may describe it at a lower level.

Exercises

1. If A and B are 4-bit signals, what kinds of logic blocks the following program will infer to ? How they may be connected?

    ```
    if (A > 5) then
        X <=   A + B ;
    else
        X <= B + 10 ;
    end if ;
    ```

2. Assume you are given a technology library containing 4-bit comparators, 4-bit adders, and 4-bit 2-to-1 multiplexers. Synthesize the following function into structure using the technology library.

    ```
    subtype INT4:  integer range 0 to 15 ;
    function TEST ( A: INT8) return INT4 is
      variable TEMP: INT4;
      if (A < 6 ) then
        TEMP <= A + 4 ;
      else
        TEMP <= 2 ;
      end if ;
      return TEMP ;
    end ;
    ```

3. For the previous problem, if you have only 2-input NAND gates in your technology library, build the circuit using minimum number of gates.

4. Given the following VHDL description, translate it into a functional equivalent hardware and draw a schematic for the circuit.

    ```
    library ieee ;
    use ieee.std_logic_1164.all ;
    ```

```
entity dontcare is
  port (X,Y    : in  std_logic ;
        OU     : out std_logic_vector(2 downto 0)) ;
end dontcare ;
architecture ex1 of dontcare is
  process test(X, Y)
  begin
    if (X='0') and (Y='0') then
      OU <= "100" ;
    elsif (X='0') and (Y='1') then
      OU <= "110" ;
    elsif (X='1') and (Y='0') then
      OU <= "011" ;
    end if ;
  end process ;
end ex1 ;
```

5. For the above problem, condition "(X='1') and (Y='1')" has not been specified. Suppose the condition will never happen. We can assign the output a random value or don't care. Compare the result of assigning output with "000" and "XXX".

6. An n-bit LFSR counter can have a maximum sequence length of $2^n - 1$. It goes through all possible code permutations except one, which is a lock-up state. The following table shows feedback connection for a shift register under 10 bits.

n	Feedback from output
3	0,1
4	0,1
5	0,2
6	0,1
7	0,1
8	0,2,3,4
9	0,4
10	0, 3

For example, a 6-bit shift register counts modulo 31, if the serial input of the register is driven by the XNOR of Q_0 and Q_1. Starting with state 000001, write the sequence of states for a 6-bit LFSR counter designed according to the table. What is the lock-up state?

7. Suppose that an n-bit LFSR counter is designed according to the table. Prove that if the even-parity circuit is changed to an odd-parity circuit, the resulting circuit is a counter that visits $2^n - 1$ states and the the lock-up state becomes 00...00.

8. Prove $Q0$ must appear on the right side of any LFSR feedback equation that generates a maximum-length sequence.

9. Translate the following program into a description at FSMD level.

```
library ieee ;
use ieee.numeric_bit ;
entity modfunc is
  port (CLK    : in  bit ;
        FLAG   : in  bit ;
        NUM    : in  unsigned(3 downto 0) ;
        NUMOU  : out unsigned(4 downto 0)) ;
end modfunc ;
architecture ex1 of dontcare is
subtype bcd4_type is unsigned(3 downto 0);
subtype bcd5_type is unsigned(4 downto 0);
constant BCD5_1    :  bcd5_type := B"0_0001";
constant BCD5_9    :  bcd5_type := B"0_1001";
function BCD_INC (L : in BCD4_TYPE) return BCD5_TYPE is
    variable V : BCD5_TYPE ;
begin
    V := L + bcd5_1 ;
    if (V > bcd5_9) then
        V := V + 6;
    end if;
    return (V) ;
end BCD_INC ;
begin
    process
      wait until CLK'event and CLK = '1';
      NUMOU <= BCD_INC(NUM) ;
      wait until CLK'event and CLK = '1';
      if (FLAG) then
```

```
            NUMOU <= BCD_INC(NUM) ;
         end if ;
      end process ;
   end ex1 ;
```

10. For the design description in the previous problem, we can have two methods to synthesize the combinational function: in-line expansion or module function. Manually synthesize the previous program using

    (a) in-line expansion method, and
    (b) module function method

    Which one gives a better result?

14

PRACTICING DESIGNS

In the next few sections, we present several examples of digital design, from a small circuit to a simple computer. Example 1 illustrates a bit clock generator which is used in serial bit transmission. In example 2, a simple vending machine is developed. In example 3, we develop circuits for a traffic light controller. Example 4 describes a design of a blackjack dealer machine and a test bench design for the machine. Finally, we demonstrate a design of a stack computer.

14.1 BIT CLOCK GENERATOR

In the transmission of serial data, bits arrive at the receiving station one at a time on a single signal line, at some nominal rate. Typical rates are 1200, 4800, 9600, or 14,400 bits per second. The data transmitter uses a clock of the appropriate frequency to regulate the serial transmission of the data bits. The receiver has no knowledge of the transmitter clock other than the agreed-upon nominal bit rate. The receiver picks off the incoming data bits by sampling the serial data line at points about midway in the interval for each data bit. To accomplish this midpoint sampling, the receiver needs a clock that delivers its active edge in the middle of each data-bit interval. The circuit to generate such a clock is called bit clock generator. Fig. 14.1 shows an input and output waveform of the circuit.

What we know about the circuit is that the preset nominal bit rate is given, and we can sense the incoming bits as voltage levels on the input line. The incoming bit stream should be sampled often enough so as not to miss any action. This is usually done by using a clock which is 16 times faster than the normal bit rate. Each time we detect a change in the voltage level on the serial

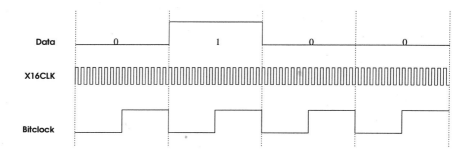

Figure 14.1 Waveform for Bit Clock Generator.

input, we will reset the bit clock to its false level. After eight clock cycles, we will change BIT_CLOCK to its true level. In the absence of a transition in the input data stream, we execute 16 clock cycles after resetting the bit clock. The following program shows a VHDL description for the bit clock generator.

```
entity BitClock is
   port ( X16CLK       : in  BIT ;
          DATA         : in  BIT ;
          RESET        : in  BIT ;
          BitClock     : out BIT ) ;
end BitClock ;
architecture Example of BitClock is
   -- mebs reset reset
   constant LEN : integer := 4 ;
   subtype Count_Type is INTEGER range 0 to (2**LEN) - 1 ;
begin
   Main : process
      variable Counter : Count_Type := 0 ;
      variable Previous_Bit : BIT ;
   begin
      wait until (X16CLK'EVENT and X16CLK = '1');
      if (Data /= Previous_Bit)or(Counter=(2**LEN)-1) then
         BitClock <= '0' ;
         Counter := 0 ;
```

```
      elsif (Counter > (2**(LEN-1)) - 1) then
         BitClock <= '1' ;
      end if ;
      Counter := (Counter + 1) mod (2 ** LEN) ;
      Previous_Bit := DATA ;
   end process Main ;
end Example ;
```

14.2 TRAFFIC LIGHT CONTROLLER

For the intersection of University Avenue (EW bound) and Campus drive (NS bound), we are going to design a traffic light controller as shown in the Fig. 14.2. Sensors raise the signal CAR_{ns} and CAR_{es} as long as a vehicle is waiting to cross the intersection.

The traffic light controller should operate as follows.

1. After reset, the E-W light should be green and the N-S light should be red.

2. When the light in one direction is red and there is no car waiting in front of the light, all lights will not change.

3. If a car is waiting in front of the red light, the circuit should check if there is any car on the other direction; if there isn't any, we can change the red light to green light; if there is, the car(s) which is(are) waiting in front of the red light should be kept waiting for a pre-set interval of time before the red light changes into green light. The light signals should remain the same until there is a car waiting in front of the (new) red light.

4. If we need to change the lights according to the rule(s) 2 and 3 stated above, we should change the lights in both directions appropriately; i.e., when we change red light into green light in one direction, we should also change the green light, in the other direction, into red light(via yellow light); there should be a reasonable delay between each change of light signal(red to green, green to yellow, and yellow to red).

The following example shows the VHDL description of a traffic light controller.

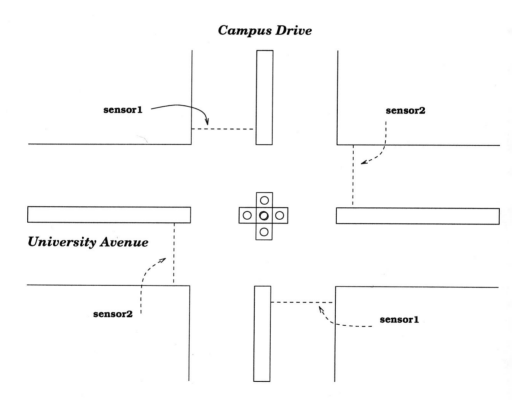

Figure 14.2 Traffic Light Controller

```
--
-- traffic light controller
--
package Traffic_Light_PKG is
   type Light is (Red, Green, Yellow) ;
   subtype Counter_Type is integer range 0 to ((2 ** 13) - 1) ;
   constant Long_Cycle   : Counter_Type := 8000 ;   -- 8 sec
   constant Midd_Cycle   : Counter_Type := 4000 ;   -- 4 sec
   constant Short_Cycle  : Counter_Type := 1000 ;   -- 1 sec
   constant Light_ON     : BIT := '0' ;             -- for LED output
   constant Light_OFF    : BIT := not Light_ON ;
end Traffic_Light_PKG ;
use WORK.TRAFFIC_LIGHT_PKG.all ;
entity TLC is
   port ( CLOCK, RESET    : in  BIT ;
          HiCAR, SideCAR  : in  BIT ;
          HiGreen         : out BIT := Light_ON ;
          HiYellow        : out BIT := Light_OFF ;
          HiRed           : out BIT := Light_OFF ;
          SiGreen         : out BIT := Light_OFF ;
          SiYellow        : out BIT := Light_OFF ;
          SiRed           : out BIT := Light_ON ) ;
end TLC ;
   -- main program
architecture Example of TLC is
begin
   Main : process
      variable Counter : Counter_Type := 0;
      -- mebs reset reset
   begin
--    highway is green, sidestreet is red
      wait on CLOCK until CLOCK = '1' and SideCAR = '1' ;
      if (HiCAR = '1') then
         Counter := 0;
         loop
```

```
               wait until CLOCK = '1' ;
               exit when (HiCAR = '0' or Counter = Long_Cycle) ;
               Counter := Counter + 1 ;
            end loop ;
         end if ;
         HiGreen     <= Light_OFF ;   SiGreen    <= Light_OFF ;
         HiYellow    <= Light_ON  ;   SiYellow   <= Light_OFF ;
         HiRed       <= Light_OFF ;   SiRed      <= Light_ON  ;
--    highway is yellow, sidestreet is red
         Counter := 0;
         loop
            wait until CLOCK = '1' ;
            exit when (Counter = Short_Cycle) ;
            Counter := Counter + 1 ;
         end loop ;
         HiGreen     <= Light_OFF ;   SiGreen    <= Light_ON  ;
         HiYellow    <= Light_OFF ;   SiYellow   <= Light_OFF ;
         HiRed       <= Light_ON  ;   SiRed      <= Light_OFF ;
--    highway is red, sidestreet is green
         wait on CLOCK until CLOCK = '1' and HiCAR = '1' ;
         if (SideCAR = '1') then
            Counter := 0;
            loop
               wait until (CLOCK = '1');
               exit when (SideCAR = '0' or Counter = Midd_Cycle);
               Counter := Counter + 1;
            end loop;
         end if;
         HiGreen     <= Light_OFF ;   SiGreen    <= Light_OFF ;
         HiYellow    <= Light_OFF ;   SiYellow   <= Light_ON  ;
         HiRed       <= Light_ON  ;   SiRed      <= Light_OFF ;
--    highway is red, sidestreet is yellow
         Counter := 1;
         loop
            wait until (CLOCK = '1') ;
```

```
            exit when (Counter = Short_Cycle) ;
            Counter := Counter + 1 ;
         end loop ;
         HiGreen    <= Light_ON  ;   SiGreen    <= Light_OFF ;
         HiYellow   <= Light_OFF ;   SiYellow   <= Light_OFF ;
         HiRed      <= Light_OFF ;   SiRed      <= Light_ON  ;
      end process ;
end Example ;
```

14.3 VENDING MACHINE

We will implement a circuit that controls a vending machine. Here is how the controller is supposed to work. A can of coke inside the machine costs 30 cents. The machine has a single coin slot that accepts nickels, dimes and quarters, one at a time. A mechanical sensor indicates to the controller whether a dime or a nickel has been put into the coin slot. The controller's output causes a can of coke to be released through the chute to the customer.

The inputs to the control are a set of three signals, that indicate what kind of coin has been deposited, as well as a reset signal. The control should generate an output signal that will cause the coke to be delivered whenever the amount of money received exceeds or equals 30 cents. If it is more than 30 cents, the circuit should return the change. After the coke has been delivered, some external circuitry will generate a reset signal to put the control back into its initial state.

To demo the circuit, we need three buttons: Q, D, and N. The *quarter* signal is asserted for one clock period when Q button is pressed; the *dime* signal is asserted for one clock period when D button is pressed; and the *nickel* signal is asserted for one clock period when N button is pressed. If the amount of money deposited is less than thirty cents, the machine displays the amount of money. If it exceeds thirty cents, the *OPEN* light is asserted and the control displays the change to be returned. Fig. 14.3 shows the block diagram of the vending machine. The following example shows the VHDL description of the vending machine.

```
-- This is a very simple vending machine design
-- You can insert a quarter, dime or nickel
```

302   CHAPTER 14

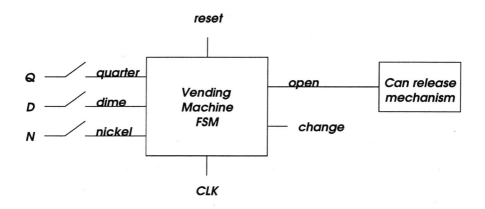

Figure 14.3 A Vending Machine Controller

```
-- When the amount of money exceed 30 cents,
-- a coke will be released
    library ieee ;
    use ieee.std_logic_1164.all ;
    entity Vending_Machine is
    port(CLK : in BIT ;
         RESET: in BIT ;
         Q:     in BIT ;
         D:     in BIT ;
         N:     in BIT ;
         Change1:  OUT STD_LOGIC_VECTOR(6 downto 0) := "1000000" ;
         Change2:  OUT STD_LOGIC_VECTOR(6 downto 0) := "1000000" ;
         Open1:    out BIT := '1') ;
    end Vending_Machine ;
    architecture Machine of Vending_Machine is
    signal Quarter, Dime, Nickel :  bit ;
    -- mebs reset_high RESET
    begin
    Q_pulse :
         process
```

```vhdl
      begin
         wait until CLK'event and CLK = '1' and Q = '1' ;
         Quarter <= '1' ;
         loop
            wait until CLK'event and CLK = '1' ;
            Quarter <= '0' ;
            exit when (Q = '0') ;
            end loop ;
         end process ;
   D_pulse :
      process
      begin
         wait until CLK'event and CLK = '1' and D = '1' ;
         Dime <= '1' ;
         loop
            wait until CLK'event and CLK = '1' ;
            Dime <= '0' ;
            exit when (D = '0') ;
            end loop ;
         end process ;
   N_pulse :
      process
      begin
         wait until CLK'event and CLK = '1' and N = '1' ;
         Nickel <= '1' ;
         loop
            wait until CLK'event and CLK = '1' ;
            Nickel <= '0' ;
            exit when (N = '0') ;
            end loop ;
         end process ;
   coke_machine :
     process
        constant RELEASE: bit := '0' ;
        constant NORELEASE: bit := '1' ;
```

```
    variable Chg :   integer range 0 to 255 := 0 ;
    variable Remn :  integer range 0 to 255 := 0 ;
begin
  wait until (CLK'event and CLK = '1' and
      ((Quarter = '1') or (Dime = '1') or (Nickel = '1'))) ;
  if(Quarter = '1') then
      Chg := Chg + 25 ;
  elsif(Dime = '1') then
      Chg := Chg + 10 ;
  elsif(Nickel = '1') then
      Chg := Chg + 5 ;
  end if;
  Remn := Chg ;
  if(Chg < 30) then
      Open1 <= NORELEASE ;
  elsif(Chg = 30) then
      Open1 <= RELEASE ;
      Chg := 0 ;
      Remn := 0 ;
  elsif(Chg > 30) then
      Open1 <= RELEASE ;
      Remn := Chg - 30 ;
      Chg := 0 ;
  end if ;
  if (Remn = 0) then
      Change1 <= "1000000" ;
      Change2 <= "1000000" ;
  elsif (Remn = 5) then
      Change1 <= "1000000" ;
      Change2 <= "0010010" ;
  elsif (Remn = 10) then
      Change1 <= "1111001" ;
      Change2 <= "1000000" ;
  elsif (Remn = 15) then
      Change1 <= "1111001" ;
```

```
                Change2 <= "0010010" ;
        elsif(Remn = 20) then
                Change1 <= "0100100" ;
                Change2 <= "1000000" ;
        elsif(Remn = 25) then
                Change1 <= "0100100" ;
                Change2 <= "0010010" ;
        elsif(Remn = 30) then
                Change1 <= "0110000" ;
                Change2 <= "1000000" ;
        else
                Change1 <= "XXXXXXX" ;
                Change2 <= "XXXXXXX" ;
        end if ;
    end process ;
end Machine ;
```

14.4 BLACK JACK DEALER MACHINE

In the next example, we are going to design a machine to simulate the dealer's actions in a blackjack game. Blackjack is a familiar card game involving a dealer and one or more players. The players can exercise their judgement but the rules specify what the dealer will do with each new card he/she receives.

14.4.1 Black Jack Dealer Design

The cards have values of 1 (ace) to 10 (10 and face cards). An ace may have the value of 1 or 11 during the play of the hand, whichever is advantageous. The dealer deals himself cards one at a time, counting ace as 11, until his/her score is greater than 16. If the dealer's score does not exceed 21, he/she "Stands", and his/her play of the hand is finished. If the dealer's score is greater than 21, he/she is "Broke" and loses the hand. The dealer must revalue an ace from 11 to 1 to avoid going broke but must then continue accepting cards ("Hits") until the count exceeds 16. The following example shows a VHDL description for the blackjack dealer.

```
        -- This is a black jack dealer machine
```

```vhdl
-- The operator first sets the input and then pushed the card_rea
button
-- The score is displayed on two seven segment displays.
library IEEE ;
use IEEE.NUMERIC_BIT.all ;
package BJ7Types_PKG is
    subtype UNSIGNED4 is UNSIGNED(3 downto 0) ;
    subtype UNSIGNED5 is UNSIGNED(4 downto 0) ;
    subtype UNSIGNED7 is UNSIGNED(6 downto 0) ;
    subtype BIT2   is INTEGER range 0 to 3 ;
    constant Light_ON  : BIT := '0' ;    -- for LED output
    constant Light_OFF : BIT := not Light_ON ;
end BJ7types_PKG ;

library IEEE;
use IEEE.NUMERIC_BIT.all ;
use WORK.BJ7types_PKG.all ;
entity BJ7SEG is
    port ( CLOCK       : in        BIT ;
           RESET       : in        BIT ;
           Card_Ready  : in        BIT ;
           Hit         : out       BIT := Light_ON  ;
           Error       : out       BIT := Light_OFF ;
           Broke       : buffer    BIT := Light_OFF ;
           Stand       : buffer    BIT := Light_OFF ;
           BlackJack   : out       BIT := Light_OFF ;
           Card_Input  : in        UNSIGNED4 ;
           Disp_Ten    : out       UNSIGNED7 ;
           Disp_One    : out       UNSIGNED7 ) ;
end BJ7SEG ;
architecture BEHAVIOR of BJ7SEG is
    signal Score       : UNSIGNED5 := "00000" ;
           --                         0
           -- 7-seg display          ---
           --                        5| 6 |1
           -- to display             ---
```

```
--                              4| 3 |2
--                                ---
-- mebs module
function decode_H4b(i:unsigned4) return unsigned7 is
    variable return_value :  unsigned7;
begin
    case (i) is
        when "0000" => return_value := "1000000" ;   -- 0
        when "0001" => return_value := "1111001" ;   -- 1
        when "0010" => return_value := "0100100" ;   -- 2
        when "0011" => return_value := "0110000" ;   -- 3
        when "0100" => return_value := "0011001" ;   -- 4
        when "0101" => return_value := "0010010" ;   -- 5
        when "0110" => return_value := "0000010" ;   -- 6
        when "0111" => return_value := "1111000" ;   -- 7
        when "1000" => return_value := "0000000" ;   -- 8
        when "1001" => return_value := "0010000" ;   -- 9
        when "1010" => return_value := "0001000" ;   -- A
        when "1011" => return_value := "0000011" ;   -- B
        when "1100" => return_value := "1000110" ;   -- C
        when "1101" => return_value := "0100001" ;   -- D
        when "1110" => return_value := "0000100" ;   -- E
        when "1111" => return_value := "0001110" ;   -- F
    end case ;
    return (return_value) ;
end decode_H4b ;
-- mebs reset reset
begin
--      black jack dealer main program
    Bj_Core : process
        variable  CARD        : UNSIGNED4 ;
        variable  T_Score     : UNSIGNED5 := "00000" ;
        variable  ACECNT      : BIT2 := 0 ;
        variable  T_Stand     : BIT ;
        variable  T_Broke     : BIT ;
```

```
begin
--      wait for a card
        wait until (rising_edge(CLOCK) and Card_Ready = '1') ;
        CARD := Card_Input ;
        if (T_Broke = '0' or T_Stand = '0') then  -- BROKE or STAND?
           T_Score := "00000" ;                    -- reset data
           ACECNT := 0 ;
        end if;
        wait until (rising_edge(CLOCK) and Card_Ready = '0');
        if (CARD > 13) then                        -- error check
           Error <= Light_On ;
        else
           ERROR <= Light_OFF ;
           if (CARD > 10) then                     -- face card check
              CARD := "01010";                     -- face card
           elsif (CARD = 1) then                   -- ace check
              T_Score := T_Score + 10 ;
              ACECNT  := ACECNT + 1 ;
           end if;
           T_Score := T_Score + CARD ;
--      status check and score adjustment
           if (T_Score < 16) then
              Hit     <= Light_On ;
              T_Stand := Light_OFF ;
              T_Broke := Light_OFF ;
           else
              while (T_Score >= 16 ) loop
                 wait until (CLOCK = '1') ;
                 HIT <= Light_OFF ;
                 if (T_Score <= 21) then
                    T_Stand := Light_ON ;
                    exit ;
                 elsif (ACECNT = 0) then
                    T_Broke := Light_ON ;
                    exit ;
```

```vhdl
                    else
                       T_Score := T_Score - 10 ;
                       ACECNT := ACECNT - 1 ;
                       if (T_Score < 16) then
                           HIT <= Light_ON ;         -- HIT after adjustment
                       end if ;
                    end if ;
                 end loop ;
            end if ;
            Score <= T_Score ;
            Stand <= T_Stand ;                       -- display on LED
            Broke <= T_Broke ;
       end if;
   end process;
--       display output
    display :   process (Score)
    begin
         if (Score = 21) then                        -- BLACK JACK
            BlackJack <= Light_ON ;
         else
            BlackJack <= Light_OFF ;
         end if ;
         if (Score >= 20) then
            Disp_Ten <= Decode_H4b("0010") ;
            Disp_One <= Decode_H4b(resize(Score-20,4)) ;
         elsif (SCORE >= 10) then
            Disp_Ten <= Decode_H4b("0001") ;
            Disp_One <= Decode_N4b(resize(Score-10,4)) ;
         else
            Disp_Ten <= Decode_N4b("0000") ;
            Disp_One <= Decode_N4b(resize(Score,4)) ;
         end if ;
    end process ;
end BEHAVIOR ;
```

14.4.2 Testbench Design

The VHDL description of the test bench for the blackjack machine is shown below. The entity declaration specifies the entity name as BJ_SIM. The absence of a port clause in this declaration indicates that there are no input or output ports for the test bench. The architecture of BJ_SIM is identified as SIMUL, and a component BJ is declared in the declarative part. The local signals in the BJ_SIM are the same as the ports of the component BJ.

The configuration specification at the end of the program associates the COMP instance of BJ to the architecture BEHAVIOR of the entity BJ in the WORK design library. In VHDL, the := symbol is used for initialization of all objects and the default initial value for the standard BIT type is '0'.

The statement part of the architecture contains four concurrent statements. It contains a component instantiation statement for the BJ component, a process to generate the waveform for the CLOCK input, a process to generate the waveform for the RESET input, and a process to read in the data and generate the waveforms for the inputs of BJ. The following shows a testbench for the balckjack dealer machine.

```
library IEEE ;
use IEEE.NUMERIC_BIT.all ;
use WORK.BJ7Types_PKG.all ;
use STD.TEXTIO.all ;
entity BJ_SIM IS end BJ_SIM ;
architecture SIMUL OF BJ_SIM IS
   component BJ
      port ( CLOCK      : in       BIT ;
             RESET      : in       BIT ;
             Card_Ready : in       BIT ;
             Hit        : out      BIT ;
             Error      : out      BIT ;
             Broke      : buffer   BIT ;
             Stand      : buffer   BIT ;
             BlackJack  : out      BIT ;
             Card_Input : in       UNSIGNED4 ;
             Disp_Ten   : out      UNSIGNED7 ;
```

```vhdl
               Disp_One    :  out        UNSIGNED7 ) ;
   end component ;
   signal CLOCK          : bit ;
   signal RESET          : bit ;
   signal Card_Ready     : bit ;
   signal Hit            : bit ;
   signal Error          : bit ;
   signal Broke          : bit ;
   signal Stand          : bit ;
   signal BlackJack      : bit ;
   signal Card_Input     : UNSIGNED4 ;
   signal Disp_Ten       : UNSIGNED7 ;
   signal Disp_One       : UNSIGNED7 ;
   signal FINISH : boolean := false ;
   file   DATAIN : TEXT is in "bj.data" ;
begin
   COMP : BJ port map(CLOCK, RESET, Card_Ready, Hit, Error,
              Broke, Stand, BlackJack, Card_Input,
                  Disp_Ten, Disp_One);
   Input : process
      variable ONELINE   : line;
      variable TEMP      : integer;
   begin
      wait until (CLOCK'event and CLOCK = '1');
      loop
         Card_Ready <= '1' ;
         readline(DATAIN, ONELINE) ;
         read(ONELINE, TEMP) ;
         Card_Input <= To_unsigned(TEMP,4) ;
         wait until (CLOCK'event and CLOCK = '1') ;
         Card_Ready <= '0' ;
         for I in 0 to 10 loop
            wait until (CLOCK'event and CLOCK = '1') ;
         end loop ;
         exit when endfile(DATAIN) ;
```

```
         end loop;
       FINISH <= true;
     end process;
     CLOCK <= not CLOCK after 100 ns;
     RRESET: process
     begin
       RESET <= '1';
       wait for 30 ns;
       RESET <= '0';
       wait until false;
     end process;
   end SIMUL;
   -- configuration declaration associates the COMP instance of BJ
to
   -- the architecture BEHAVIOR of the entity BJ in the WORK library
   configuration BJ_BE_CONF of BJ_SIM is
     for SIMUL
       for COMP : BJ use entity WORK.BJ(BEHAVIOR);
       end for;
     end for;
   end BJ_BE_CONF;
```

14.5 DESIGNING A STACK COMPUTER

We will implement a stack computer in this section. The machine is organized around an evaluation stack. It supports 16 different instructions, including those for accessing memory, data manipulation, and conditional/unconditional branches. These instructions are encoded in one to three 4-bit words. The machine can address 16 four-bit words.

A stack is a data structure in which the element last added is the first to be deleted. Hence, stacks are often called "last in first out" (LIFO) data structures. Items are PUSHed to the top of the stack and POPed from the top of the stack. Consider the expression 9-(5+1). This can be implemented by the following sequence of stack-oriented operations:

PUSH 9		push constant 9 to STR of stack
PUSH 5		push constant 5 to STR of stack
PUSH 1		push constant 1 to STR of stack
ADD		add top two elements of the stack, remove and replace with the number 7
SUB		top two elements of the stack are subtracted, remove and replace with the number 2

Instructions are encoded from one to three four-bit words. Arithmetic, logical, and shift instructions are encoded in a single word: all four bits of the op code. The operands are implicitly the elements on the top of the stack. The arithmetic/logic instructions are: ADD, SUB, ANDL, ORL, COMP, INCR, RSR (rotation shift), ASR (arithmetic shift), and their encodings are the following:

0000	ADD	MEM(STR-1) := MEM(STR-1) + MEM(STR); STR := STR -1;
0001	SUB	MEM(STR-1) := MEM(STR-1) - MEM(STR); STR := STR -1;
0010	ANDL	MEM(STR-1) := MEM(STR-1) and MEM(STR); STR := STR -1;
0011	ORL	MEM(STR-1) := MEM(STR-1) or MEM(STR); STR := STR -1;
0100	COMP	MEM(STR) := not MEM(STR);
0101	INCR	MEM(STR) := MEM(STR) + 1;
0110	RSR	MEM(STR) := MEM(STR)(0) & MEM(STR)(3 downto 1) ;
0111	ASR	MEM(STR) := MEM(STR)(3) & MEM(STR)(3 downto 1) ;

The next instruction places data onto the stack or removes data from the stack. The PUSH instruction is encoded in two 4-bit words, one that contains the op code and one that contains 4-bit twos complement data to place on the top of the stack.

1000 $X_3 X_2 X_1 X_0$ PUSH data   STR := STR + 1;

The next instruction group are are conditional/unconditional branches. All are encoded in three 4-bits words: one word for the op code, and one word for the target address. The conditional branch instructions, BRZ and BRN, test the top of stack element for $= 0$ or < 0, respectively. If true, the PC is changed to the target address. In either case, the top element is left undisturbed.

1001 $A_3A_2A_1A_0$	BRZ address	IF MEM(STR) = 0 then $\quad$ PC := $A_3A_2A_1A_0$;
1010 $A_3A_2A_1A_0$	BRN address	IF MEM(STR)(3) = '1' then $\quad$ PC := $A_3A_2A_1A_0$;
1011 $A_3A_2A_1A_0$	JMP address	PC := $A_3A_2A_1A_0$;

The next instruction group are subroutine call and return. JSR (Jump to Subroutine) is a special instruction that is used to implement subroutines. The current value of the PC is placed on the stack and then the PC is changed to the target address. Any values placed on the stack by the subroutine must be POPed before it returns. The RTS instruction restores the PC from the value saved on the stack.

1100 $A_3A_2A_1A_0$	JSR address	STR := STR + 1; MEM[STR] := PC; PC := $A_3A_2A_1A_0$
1101	RTS	PC := MEM[STR]; STR := STR - 1;

The last instruction is to display the element on the top of the stack to a seven-segment display.

1110	DISP	7-seg := MEM(STR);
1111	HALT	stop the program execution;

The following example shows a VHDL description for the Stack Computer design.

```
library ieee;
use ieee.numeric_bit.all;
```

```vhdl
entity Computer is
port(clock :   in bit;
     reset :   in bit;
     button :  in bit;
     seg   :   out unsigned(6 downto 0));
end Computer;
architecture Stack of Computer is
  -- mebs reset_high reset
  subtype bit4 is integer range 0 to 15;
  subtype unsigned4 is unsigned(3 downto 0);
  subtype unsigned7 is unsigned(6 downto 0);
  type Memory is array (bit4) of unsigned(3 downto 0);
  --mebs module
  function decode_H4b(i:unsigned4) return unsigned7 is
    variable return_value :  unsigned7;
  begin
    case (i) is
      when "0000" => return_value := "1000000";   -- 0
      when "0001" => return_value := "1111001";   -- 1
      when "0010" => return_value := "0100100";   -- 2
      when "0011" => return_value := "0110000";   -- 3
      when "0100" => return_value := "0011001";   -- 4
      when "0101" => return_value := "0010010";   -- 5
      when "0110" => return_value := "0000010";   -- 6
      when "0111" => return_value := "1111000";   -- 7
      when "1000" => return_value := "0000000";   -- 8
      when "1001" => return_value := "0010000";   -- 9
      when "1010" => return_value := "0001000";   -- A
      when "1011" => return_value := "0000011";   -- B
      when "1100" => return_value := "1000110";   -- C
      when "1101" => return_value := "0100001";   -- D
      when "1110" => return_value := "0000100";   -- E
      when "1111" => return_value := "0001110";   -- F
    end case ;
    return(return_value);
```

```
     end;
    ----------------------------------------------------------
    --Variables used:
    -- PC  : Program counter
    -- STR :Pointer to top of stack
    -- STR_1 :Pointer to second top of stack
    -- ACC :New instruction from the program
    -- opa :  Operand to store
    ----------------------------------------------------------
    begin
    main : process
      constant ADD    : unsigned4 := "0000";
      constant SUB    : unsigned4 := "0001";
      constant ANDL   : unsigned4 := "0010";
      constant ORL    : unsigned4 := "0011";
      constant COMP   : unsigned4 := "0100";
      constant INCR   : unsigned4 := "0101";
      constant RSR    : unsigned4 := "0110";
      constant ASR    : unsigned4 := "0111";
      constant PUSH   : unsigned4 := "1000";
      constant BRZ    : unsigned4 := "1001";
      constant BRN    : unsigned4 := "1010";
      constant JMP    : unsigned4 := "1011";
      constant JSR    : unsigned4 := "1100";
      constant RTS    : unsigned4 := "1101";
      constant DISP   : unsigned4 := "1110";
      constant HALT   : unsigned4 := "1111";
      constant Program :Memory :=
        ( PUSH,"1001",         -- push 9
          DISP,                -- disp
          PUSH,"0101",         -- push 5
          DISP,                -- disp
          ADD ,                -- add
          DISP,                -- disp
          PUSH,"0001",         -- push 1
```

```
              DISP,                   -- disp
              SUB ,                   -- sub
              DISP,                   -- disp
              HALT,                   -- halt
              HALT,
              HALT );
   variable stack   :Memory;
   variable PC,STR  :bit4 := 0;
   variable STR_1   :bit4;
   variable ACC         :unsigned4;
   variable opa         :unsigned4;
   variable memaddr :bit4;
begin
   wait until (clock'event and clock = '1');
   ACC := Program(PC);
   PC := PC + 1;
   case (ACC) is
   --ADD instruction
   when ADD =>
       STR_1 := STR - 1;
       stack(STR_1) := (stack(STR_1) + stack(STR)) mod 16;
       STR := STR_1;
   --SUB (Subtract)
   when SUB =>
       STR_1 := STR - 1;
       stack(STR_1) := (stack(STR_1) - stack(STR)) mod 16;
       STR := STR_1;
   --AND
   when ANDL =>
       STR_1 := STR - 1;
       stack(STR_1) := stack(STR_1) and stack(STR);
       STR := STR_1;
   --OR
   when ORL =>
       STR_1 := STR - 1;
```

```
        stack(STR_1) := stack(STR_1) or stack(STR);
        STR := STR_1;
--COMP (Complement)
when COMP =>
        stack(STR) := not stack(STR);
--INCR (Increment)
when INCR =>
        stack(STR) := (stack(STR) + 1) mod 16;
--RSR(Right shift)
when RSR =>
        opa := stack(STR);
        opa := opa(0) & opa(3 downto 1);
        stack(STR) := opa;
--ASR(Arithmetic Right shift)
when ASR =>
        opa := stack(STR);
        opa := opa(3) & opa(3 downto 1);
        stack(STR) := opa;
--PUSH (To push data)
when PUSH =>
        STR := STR + 1;
        stack(STR) := Program(PC);
        PC := PC + 1;
--BRZ (Branch on 0)
when BRZ =>
        if (stack(STR) = 0) then
           PC := to_integer(Program(PC));
        else
           PC := PC + 1;
        end if;
--BRN (Branch on not 0)
when BRN =>
        opa := stack(STR);
        if (opa(3) = '1') then
            PC := to_integer(Program(PC));
```

```
            else
                PC := PC + 1;
            end if;
    --JMP (Jump to a location)
    when JMP =>
            PC := to_integer(Program(PC));
    --JSR (Jump tp subroutine)
    when JSR =>
            STR := STR + 1;
            stack(STR) := to_unsigned(PC + 1,4);
            PC := to_integer(Program(PC));
    --RTS (Return to caller)
    when RST =>
            PC := stack(STR);
            STR := STR - 1;
    --DISP (Display top of stack)
    when DISP =>
            seg <= decode_H4b(stack(STR));
    --HALT (Halt the program)
    when HALT =>
            wait until clock'event and clock='1' and button = '1';
    end case;
end process;
end Stack;
```

Exercises

1. An 8-bit parallel-to-serial converter circuit is to be designed. The circuit remains in an idle state as long as the START input is false. But when this START input becomes true, the 8-bit data BYTE is loaded into the shift register and the right-shifting of the data begins. After the 8 bits are shifted out, the circuit returns to the idle state. Write an algorithmic level description for the circuit.

2. For the transmission of asynchronous serial data, a start bit must be inserted before each data byte and s stop bit inserted after the data byte. Also, for error-checking purposes, sometimes a parity bit is inserted between the data byte and the stop bit. In this problem, you are to redesign the parallel-to-serial converter so that it will perform these insertions. Specifically, when the START input is false, the converter is to be in an idle state in which the output is high. But when START becomes true, the converter loads the 8-bit data into the shift register and shifts out the start bit (low) first, after which it begins the shifting out the parity bit, followed by the stop bit.

3. A lock for a car can be opened with a correct combination or with a key. If the combination feature is used, then the clock must be supplied with the correct combination within three attempts. Otherwise an alarm will sound. Specifications for the lock circuit are as follows:

    (a) After the first unsuccessful attempt, the circuit outputs signal MSG1,

    (b) After the second unsuccessful attempt, output signal MSG2 and wait for some time before being ready to be tried again (READY).

    (c) After the third successive unsuccessful attempt, sound an alarm.

    (d) When the car is opened, reset to allow three more tries.

    (e) When the key is used, stop the alarm and reset to allow three more tries.

    What are the input and output signals of the circuit? Write an algorithmic design description for the circuit.

4. A traffic light controller is to be designed that will operate a traffic light at the intersection of a main highway and a less frequently used farm road. Traffic sensors are placed on both the highway and the farm road to indicate when traffic is present. If no traffic is sensed on the farm road, traffic on the highway is allowed to flow. But when a vehicle on the farm

road, the vehicle on the farm road must wait for 30 seconds or until the high way is clear, whichever occurs first. Once the farm road vehicle has gone, the system must permit traffic to resume on the high way. Write an algorithmic design description for the circuit.

5. Consider a simple computer with a main memory M having a capacity of 2^{14} 16-bit words. The CPU contains an n-bit accumulator AC and an n-bit program counter PC. Instructions for a single accumulator machine are called single address instructions. This is because there is only single reference to the memory. One operand is implicitly the AC, and the other is an operand in the memory. The instructions and data words are 16 bits. The higher-order bits of the instruction contain an operation code to denote the operation type. The remaining 14-bits are used as the memory address of the operand word. The computer supports four instructions:

   (a) Load from memory
   LOD X: $M[X] \Rightarrow AC$;

   (b) Store to memory
   STR X: $AC \Rightarrow M[X]$;

   (c) Add from memory
   ADD X: $AC + M[X] \Rightarrow AC$;

   (d) Branch if AC negative
   BRN X: if $AC<15>=1$ then $X \Rightarrow PC$;

   Write an algorithmic design description for the circuit.

REFERENCES

[1] R. Lipsett, C. Schaefer, and C. Ussery, *VHDL: Hardware Description and Design,* Kluwer Academic Publishers, 1989.

[2] D. Perry, *VHDL,* McGraw-Hill, 1994.

[3] Steve Carlson, *Introduction to HDL-based Design using VHDL,* Synopsys, 1991.

[4] Zainalabedin Navabi, *VHDL, Analysis and Modeling of Digital Systems,* Prentice-Hall, 1993.

[5] Stanley Mazor and Patricia Langstraat, *A Guide To VHDL,* Kluwer Academic Publishers, 1992.

[6] *HDL Synthesis,* Reference Manual, Exemplar Logic, Inc., 1994.

[7] *VHDL Compiler References,* Synopsys, Inc., 1990.

[8] *IEEE Standard 1076-1987 VHDL Language Reference Manual,* IEEE, 1987.

[9] *IEEE Standard 1076-1993 VHDL Language Reference Manual,* IEEE, 1994.

[10] *Behavior Compiler Methodology Manual,* Synopsys, Inc., 1995.

[11] James R. Armstrong, *Chip-Level Modeling with VHDL* Prentice Hall, 1989.

[12] D.D. Gajski and R. Kuhn, "Guest Editors' Introduction: New VLSI Tools," *IEEE Computer,* vol. 16, no. 12, pp. 11-14, Dec. 1983.

[13] John Wakerly, *Digital Design Principles and Practices,* Prentice-Hall, 1990.

[14] Randy Katz, *Comtemporary Logic Design,* The Benjamin/Cummings Publishing Company, 1994.

[15] Ronald J. Tocci, *Digital Systems, principles and applications*, Prentice-Hall Inc., 1991.

[16] Herman Lam and John O'malley, *Fundamentals of Computer Engineering*, John Wiley & Sons, Inc., 1988.

[17] Richard Tinder, *Digital Engineering Design*, Prentice-Hall Inc., 1991.

[18] Franklin Prosser and David Winkel, *The Art of Digital Design*, Prentice-Hall Inc., 1987.

[19] V. Nelson, H. Nagle, B. Carroll, and J. Irwin, *Digital Logic Circuit Analysis & Design*, Prentice Hall, 1995,

[20] R. Camposano, and Wayne Wolf, *High-Level VLSI Synthesis*, Kluwer Academic Publishers, 1991.

[21] D. Gajski, N. Dutt, A. Wu, and S. Lin, *High-Level Synthesis*, Kluwer Academic Publishers, 1992.

[22] R.A. Bergamaschi, and A. Kuehlmann, "*A System for Production Use of High-Level Synthesis*," IEEE Trans. on VLSI Systems, pp.233-243, September 1993.

[23] J. Biesenack, M. Koster, A. Langmaier, S. Ledeux, S. Marz, M. Payer, M. Pilsl, S. Rumler, H. Soukup, N. Wehn, and P. Duzy, "*The Siemens High-Level Synthesis System CALLAS*," IEEE Trans. on VLSI Systems, pp.244-251, September 1993.

[24] R. Camposano, "*Path-based Scheduling for Synthesis*," IEEE Trans. on Computer Aided Design, pp. 85-93, Jan. 1991.

[25] K. O'Brien, M. Rahmouni, and Jerraya, "*A VHDL-Based Scheduling Algorithm for Control Flow Dominated Circuits*," Proc. of High Level Synthesis Workshop, pp.135-145, 1992.

[26] G. De Micheli et al., "*The Olympus Synthesis System for Digital Design*," IEEE Design and Test of Computers, pp. 37-53, Oct. 1990.

[27] Wayne Wolf, Andres Takach, Chun-Yao Huang, Richard Manno and Ephrem Wu, "*The Princeton University Behavior Synthesis System*," 29th Design Automation Conference, pp. 182-187, June, 1992.

[28] *IEEE Standard VHDL Language Reference Manual*, March 1988. (IEEE Std 1076-1987)

REFERENCES

[29] Frank Vahid and Daniel D. Gajski, *"Specification Partitioning for System Design,"* 29th Design Automation Conference, June, 1992.

[30] S. Narayan, F. Vahid, and D.D. Gajski, *"System specification and Synthesis with the SpecCharts Language,"* Proc. of ICCAD-91, pp.266-269, IEEE, 1991.

[31] E.D. Lagnese and D.E. Thomos, *"Architectural partitioning for System Level Design,"* 26th Design Automation Conference, June, 1989.

[32] Rajesh Gupta, Claudionor Coelho and Giovanni De Micheli, *"Synthesis and Simulation of Digital Systems Containing Interacting Hardware and Software Components,"* 29th Design Automation Conference, pp. 225-230, June, 1992.

[33] A.V. Aho, R. Sethi, and J.D. Ullman, "Compilers: Principles, Techniques, and Tools", *Addison Wesley*, 1986.

[34] J.A. Fisher, "Trace Scheduling: A Technique for Global Microcode Compaction", *IEEE Transaction on Computers*, pp.478-490, June 1981.

[35] D. Brand and V. Iyengar. "Timing analysis using functional analysis". *IEEE Trans. on Computers*, pages 1309–1314, October 1988.

[36] T. Kim, J. Liu and C.L. Liu. *"A scheduling algorithm for conditional resource sharing,"* Proc. of International Conference on Computer-Aided Design, 1991, pp. 84-87.

[37] C.Y. Roger Chen and Michael Z. Moricz, *"Data path scheduling for two-level pipelining,"* Proc. ACM/IEEE 28th Design Automation Conf., 1991, pp. 603-606.

[38] C. T. Hwang, Y. C. Hsu and Y. L. Lin, "PLS: A Scheduler for Pipeline Synthesis", *IEEE Transactions on Computer-Aided Design of Integrated Circuits and Systems,* pp. 1279-1286, Sep., 1993.

[39] C.T. Hwang, J.H. Lee and Y.C. Hsu, "A Formal Approach to the Scheduling Problem in High Level Synthesis", *IEEE Transactions on Computer-Aided Design of Integrated Circuits and Systems,* pp. 464-475, April, 1991.

[40] Y. C. Hsu, T. Y. Liu, F. S. Tsai, S. Z. Lin and C. Yu, *"Digital Design from Concept to Prototype in Hours,"* APCCAS, Dec. 1994.

[41] F. S. Tsai and Y. C. Hsu, *"STAR: A System for Hardware Allocation in Data Path Synthesis,"* IEEE Transactions on Computer-Aided Design of Integrated Circuits and Systems, pp. 1053-1064, Sep., 1992.

[42] Tak Yin Wang, Ta-Yung Liu and Yu-Chin Hsu, "*Synthesis of Datapath Modules in MEBS,*" Technical Report, U.C. Riverside, 1994.

[43] J. R. Burch, E. M. Clarke, K.L. McMillan and D.L. Dill, "*Sequential circuit Verification using Symbolic Model Checking,*" 27th Design Automation Conference, June, 1990.

[44] R. Brayton, R. Rudell, A. Sangiovanni-Vincentelli, and A. Wang, "*MIS: A Multiple-level Logic Optimization System,*" IEEE Trans. on Computer-Aided Design of Integrated Circuits and Systems, pp. 1062-1081, Sep., 1987.

[45] *XACT Development System, Libraries Guide,* January 1993.

[46] *V-System/Workstation User's Manual,* Version 4, Model Technology, July 1993.

[47] *EDIF Electronic Design Interchange Format Version 2 0 0* , Electronic Industries Association, 1989.

[48] *Xilinx XNF Netlist Specification,* Xilinx Inc. Version 5.00, 1993.

A
RESERVED WORDS

Reserved words are identifiers that are reserved for the use in the language. Each of them has a fixed meaning and may not be used for other purposes in VHDL. An identifier must begin with an alphabetic letter (a - z) followed by letters, underscores, or digits. VHDL identifiers are *not case-sensitive*. In other words, upper and lower case letters are considered as being the same.

abs	access	after	alias
all	and	architecture	array
assert	attribute	begin	block
body	buffer	bus	case
component	configuration	constant	disconnect
downto	else	elsif	end
entity	exit	file	for
function	generate	generic	guarded
if	in	inout	is
label	library	linkage	loop
map	mod	nand	new
next	nor	null	of
on	open	or	others
out	package	port	procedure
process	range	record	register
rem	report	return	select
severity	signal	subtype	then
to	transport	type	units
util	use	variable	wait

when		while		with		xor

B

STANDARD LIBRARY PACKAGES

IEEE standard committee defines two libraries for VHDL, namely STD and IEEE, each contains several packages

Library STD contains two packages: STANDARD and TEXTIO. The STANDARD package (1076-1987) defines some useful data types, such as INTEGER, BIT, BOOLEAN, *etc* that are automatically visible in any VHDL description. The TEXTIO package defines types and operations for communication with a standard programming environment. This TEXTIO package is needed for simulation purposes.

Library IEEE contains an Standard Logic package (STD_LOGIC_1164) and Standard Synthesis packages (NUMERIC_BIT and NUMERIC_STD – pending for approval). These three packages contain some useful types and functions that are required for VHDL synthesis.

B.1 THE STANDARD PACKAGE

The package STANDARD is implicitly used by all design entities. Its source is defined as follows.[1]

```
------------------------------------------------------------------
-- This is Package STANDARD as defined in the VHDL Language Reference Manual.
------------------------------------------------------------------
package standard is
```

---
[1] Copyright ©1994 IEEE. All Rights Reserved.

```
type boolean is (false,true);
type bit is ('0', '1');
type character is (
        nul, soh, stx, etx, eot, enq, ack, bel,
        bs,  ht,  lf,  vt,  ff,  cr,  so,  si,
        dle, dc1, dc2, dc3, dc4, nak, syn, etb,
        can, em,  sub, esc, fsp, gsp, rsp, usp,
        ' ', '!', '"', '#', '$', '%', '&', ''',
        '(', ')', '*', '+', ',', '-', '.', '/',
        '0', '1', '2', '3', '4', '5', '6', '7',
        '8', '9', ':', ';', '<', '=', '>', '?',
        '@', 'A', 'B', 'C', 'D', 'E', 'F', 'G',
        'H', 'I', 'J', 'K', 'L', 'M', 'N', 'O',
        'P', 'Q', 'R', 'S', 'T', 'U', 'V', 'W',
        'X', 'Y', 'Z', '[', '\', ']', '^', '_',
        '`', 'a', 'b', 'c', 'd', 'e', 'f', 'g',
        'h', 'i', 'j', 'k', 'l', 'm', 'n', 'o',
        'p', 'q', 'r', 's', 't', 'u', 'v', 'w',
        'x', 'y', 'z', '{', '|', '}', '~', del);
type severity_level is (note, warning, error, failure);
type integer is range -2147483648 to 2147483647;
type real is range -1.0E38 to 1.0E38;
type time is range -2147483647 to 2147483647
   units
            fs;
            ps = 1000 fs;
            ns = 1000 ps;
            us = 1000 ns;
            ms = 1000 us;
            sec = 1000 ms;
            min = 60 sec;
            hr = 60 min;
   end units;
function now return time;
subtype natural is integer range 0 to integer'high;
subtype positive is integer range 1 to integer'high;
```

```
            type string is array (positive range <>) of character;
            type bit_vector is array (natural range <>) of bit;
end standard;
```

B.2 THE TEXTIO PACKAGE

The package TEXTIO defines a number of types and functions which are used to read and write ASCII files. The package declaration for TEXTIO is given below: [2]

```
-------------------------------------------------------------------------------
-- Package TEXTIO as defined in Chapter 14 of the IEEE Standard VHDL
--    Language Reference Manual (IEEE Std.  1076-1987)
-------------------------------------------------------------------------------
package TEXTIO is
    type LINE is access string;
    type TEXT is file of string;
    type SIDE is (right, left);
    subtype WIDTH is natural;
    file input :   TEXT is in "STD_INPUT";
    file output :  TEXT is out "STD_OUTPUT";
    procedure READLINE(variable f:in TEXT; L: inout LINE);
    procedure READ(L:inout LINE; VALUE: out bit; GOOD : out BOOLEAN);
    procedure READ(L:inout LINE; VALUE: out bit);
    procedure READ(L:inout LINE; VALUE: out bit_vector; GOOD : out BOOLEAN);
    procedure READ(L:inout LINE; VALUE: out bit_vector);
    procedure READ(L:inout LINE; VALUE: out BOOLEAN; GOOD : out BOOLEAN);
    procedure READ(L:inout LINE; VALUE: out BOOLEAN);
    procedure READ(L:inout LINE; VALUE: out character; GOOD : out BOOLEAN);
    procedure READ(L:inout LINE; VALUE: out character);
    procedure READ(L:inout LINE; VALUE: out integer; GOOD : out BOOLEAN);
    procedure READ(L:inout LINE; VALUE: out integer);
    procedure READ(L:inout LINE; VALUE: out real; GOOD : out BOOLEAN);
```

---
[2]Copyright ©1994 IEEE. All Rights Reserved.

```
procedure READ(L:inout LINE; VALUE: out real);
procedure READ(L:inout LINE; VALUE: out string; GOOD : out BOOLEAN);
procedure READ(L:inout LINE; VALUE: out string);
procedure READ(L:inout LINE; VALUE: out time; GOOD : out BOOLEAN);
procedure READ(L:inout LINE; VALUE: out time);
procedure WRITELINE(f :   out TEXT; L : inout LINE);
procedure WRITE(L : inout LINE; VALUE : in bit;
   JUSTIFIED: in SIDE := right;
   FIELD: in WIDTH := 0);
procedure WRITE(L : inout LINE; VALUE : in bit_vector;
   JUSTIFIED: in SIDE := right;
   FIELD: in WIDTH := 0);
procedure WRITE(L : inout LINE; VALUE : in BOOLEAN;
   JUSTIFIED: in SIDE := right;
   FIELD: in WIDTH := 0);
procedure WRITE(L : inout LINE; VALUE : in character;
   JUSTIFIED: in SIDE := right;
   FIELD: in WIDTH := 0);
procedure WRITE(L : inout LINE; VALUE : in integer;
   JUSTIFIED: in SIDE := right;
   FIELD: in WIDTH := 0);
procedure WRITE(L : inout LINE; VALUE : in real;
   JUSTIFIED: in SIDE := right;
   FIELD: in WIDTH := 0;
   DIGITS: in NATURAL := 0);
procedure WRITE(L : inout LINE; VALUE : in string;
   JUSTIFIED: in SIDE := right;
   FIELD: in WIDTH := 0);
procedure WRITE(L : inout LINE; VALUE : in time;
   JUSTIFIED: in SIDE := right;
   FIELD: in WIDTH := 0;
   UNIT: in TIME := ns);
end;
```

B.3 THE STANDARD LOGIC PACKAGE

The package declaration of the STD_LOGIC_1164 is listed here for reference purpose[3]

```
-- -----------------------------------------------------------------
--
--   Title      :  std_logic_1164 multi-value logic system
--   Library    :  This package shall be compiled into a library
--              :  symbolically named IEEE.
--              :
--   Developers:   IEEE model standards group (par 1164)
--   Purpose    :  This packages defines a standard for designers
--              :  to use in describing the interconnection data types
--              :  used in vhdl modeling.
--              :
--   Limitation:   The logic system defined in this package may
--              :  be insufficient for modeling switched transistors,
--              :  since such a requirement is out of the scope of this
--              :  effort.  Furthermore, mathematics, primitives,
--              :  timing standards, etc. are considered orthogonal
--              :  issues as it relates to this package and are therefore
--              :  beyond the scope of this effort.
--              :
--   Note       :  No declarations or definitions shall be included in,
--              :  or excluded from this package.  The "package declaration"
--              :  defines the types, subtypes and declarations of
--              :  std_logic_1164.  The std_logic_1164 package body shall be
--              :  considered the formal definition of the semantics of
--              :  this package.  Tool developers may choose to implement
--              :  the package body in the most efficient manner available
--              :  to them.
--              :
-- -----------------------------------------------------------------
--   modification history :
```

---

[3]Copyright ©1993 IEEE. All Rights Reserved.

```
-- ---------------------------------------------------------------
-- version | mod. date:|
--  v4.200 | 01/02/92  |
-- ---------------------------------------------------------------
PACKAGE std_logic_1164 IS
    -------------------------------------------------------------
    -- logic state system   (unresolved)
    -------------------------------------------------------------
    TYPE std_ulogic IS ( 'U',   -- Uninitialized
                         'X',   -- Forcing  Unknown
                         '0',   -- Forcing  0
                         '1',   -- Forcing  1
                         'Z',   -- High Impedance
                         'W',   -- Weak     Unknown
                         'L',   -- Weak     0
                         'H',   -- Weak     1
                         '-'    -- Don't care
                       );
    -------------------------------------------------------------
    -- unconstrained array of std_ulogic for use with the resolution function
    -------------------------------------------------------------
    TYPE std_ulogic_vector IS ARRAY ( NATURAL RANGE <> ) OF std_ulogic;
    -------------------------------------------------------------
    -- resolution function
    -------------------------------------------------------------
    FUNCTION resolved ( s : std_ulogic_vector ) RETURN std_ulogic;
    -------------------------------------------------------------
    -- *** industry standard logic type ***
    -------------------------------------------------------------
    SUBTYPE std_logic IS resolved std_ulogic;
    -------------------------------------------------------------
    -- unconstrained array of std_logic for use in declaring signal arrays
    -------------------------------------------------------------
    TYPE std_logic_vector IS ARRAY ( NATURAL RANGE <>) OF std_logic;
    -------------------------------------------------------------
```

```
    -- common subtypes
    ----------------------------------------------------------------
    SUBTYPE X01     IS resolved std_ulogic RANGE 'X' TO '1';
    SUBTYPE X01Z    IS resolved std_ulogic RANGE 'X' TO 'Z';
    SUBTYPE UX01    IS resolved std_ulogic RANGE 'U' TO '1';
    SUBTYPE UX01Z   IS resolved std_ulogic RANGE 'U' TO 'Z';
    ----------------------------------------------------------------
    -- overloaded logical operators
    ----------------------------------------------------------------
    FUNCTION "and"  ( l :   std_ulogic; r :   std_ulogic ) RETURN UX01;
    FUNCTION "nand" ( l :   std_ulogic; r :   std_ulogic ) RETURN UX01;
    FUNCTION "or"   ( l :   std_ulogic; r :   std_ulogic ) RETURN UX01;
    FUNCTION "nor"  ( l :   std_ulogic; r :   std_ulogic ) RETURN UX01;
    FUNCTION "xor"  ( l :   std_ulogic; r :   std_ulogic ) RETURN UX01;
--  function "xnor" ( l :   std_ulogic; r :   std_ulogic ) return ux01;
    FUNCTION "not"  ( l :   std_ulogic                   ) RETURN UX01;
    ----------------------------------------------------------------
    -- vectorized overloaded logical operators
    ----------------------------------------------------------------
    FUNCTION "and"  ( l, r : std_logic_vector  ) RETURN std_logic_vector;
    FUNCTION "and"  ( l, r : std_ulogic_vector ) RETURN std_ulogic_vector;
    FUNCTION "nand" ( l, r : std_logic_vector  ) RETURN std_logic_vector;
    FUNCTION "nand" ( l, r : std_ulogic_vector ) RETURN std_ulogic_vector;
    FUNCTION "or"   ( l, r : std_logic_vector  ) RETURN std_logic_vector;
    FUNCTION "or"   ( l, r : std_ulogic_vector ) RETURN std_ulogic_vector;
    FUNCTION "nor"  ( l, r : std_logic_vector  ) RETURN std_logic_vector;
    FUNCTION "nor"  ( l, r : std_ulogic_vector ) RETURN std_ulogic_vector;
    FUNCTION "xor"  ( l, r : std_logic_vector  ) RETURN std_logic_vector;
    FUNCTION "xor"  ( l, r : std_ulogic_vector ) RETURN std_ulogic_vector;
--  ----------------------------------------------------------------
--  Note : The declaration and implementation of the "xnor" function is
--  specifically commented until at which time the VHDL language has been
--  officially adopted as containing such a function. At such a point,
--  the following comments may be removed along with this notice without
--  further "official" balloting of this std_logic_1164 package. It is
```

```
--    the intent of this effort to provide such a function once it becomes
--    available in the VHDL standard.
--    ------------------------------------------------------------------
--    function "xnor" ( l, r :   std_logic_vector  ) return std_logic_vector;
--    function "xnor" ( l, r :   std_ulogic_vector ) return std_ulogic_vector;
      FUNCTION "not"  ( l :  std_logic_vector  ) RETURN std_logic_vector;
      FUNCTION "not"  ( l :  std_ulogic_vector ) RETURN std_ulogic_vector;
      ------------------------------------------------------------------
      -- conversion functions
      ------------------------------------------------------------------
      FUNCTION To_bit       ( s :  std_ulogic;           xmap : BIT := '0') RETURN BIT;
      FUNCTION To_bitvector ( s :  std_logic_vector ; xmap : BIT := '0') RETURN BIT_VECTOR;
      FUNCTION To_bitvector ( s :  std_ulogic_vector; xmap : BIT := '0') RETURN BIT_VECTOR;
      FUNCTION To_StdULogic       ( b : BIT              ) RETURN std_ulogic;
      FUNCTION To_StdLogicVector  ( b : BIT_VECTOR       ) RETURN std_logic_vector;
      FUNCTION To_StdLogicVector  ( s : std_ulogic_vector ) RETURN std_logic_vector;
      FUNCTION To_StdULogicVector ( b : BIT_VECTOR       ) RETURN std_ulogic_vector
      FUNCTION To_StdULogicVector ( s : std_logic_vector ) RETURN std_ulogic_vector
      ------------------------------------------------------------------
      -- strength strippers and type convertors
      ------------------------------------------------------------------
      FUNCTION To_X01  ( s :  std_logic_vector  ) RETURN  std_logic_vector;
      FUNCTION To_X01  ( s :  std_ulogic_vector ) RETURN  std_ulogic_vector;
      FUNCTION To_X01  ( s :  std_ulogic        ) RETURN  X01;
      FUNCTION To_X01  ( b :  BIT_VECTOR        ) RETURN  std_logic_vector;
      FUNCTION To_X01  ( b :  BIT_VECTOR        ) RETURN  std_ulogic_vector;
      FUNCTION To_X01  ( b :  BIT               ) RETURN  X01;
      FUNCTION To_X01Z ( s :  std_logic_vector  ) RETURN  std_logic_vector;
      FUNCTION To_X01Z ( s :  std_ulogic_vector ) RETURN  std_ulogic_vector;
      FUNCTION To_X01Z ( s :  std_ulogic        ) RETURN  X01Z;
      FUNCTION To_X01Z ( b :  BIT_VECTOR        ) RETURN  std_logic_vector;
      FUNCTION To_X01Z ( b :  BIT_VECTOR        ) RETURN  std_ulogic_vector;
```

```
    FUNCTION To_X01Z ( b :  BIT              ) RETURN  X01Z;
    FUNCTION To_UX01 ( s :  std_logic_vector ) RETURN  std_logic_vector;
    FUNCTION To_UX01 ( s :  std_ulogic_vector) RETURN  std_ulogic_vector;
    FUNCTION To_UX01 ( s :  std_ulogic       ) RETURN  UX01;
    FUNCTION To_UX01 ( b :  BIT_VECTOR       ) RETURN  std_logic_vector;
    FUNCTION To_UX01 ( b :  BIT_VECTOR       ) RETURN  std_ulogic_vector;
    FUNCTION To_UX01 ( b :  BIT              ) RETURN  UX01;
    ------------------------------------------------------------------
    -- edge detection
    ------------------------------------------------------------------
    FUNCTION rising_edge  (SIGNAL s :  std_ulogic) RETURN BOOLEAN;
    FUNCTION falling_edge (SIGNAL s :  std_ulogic) RETURN BOOLEAN;
    ------------------------------------------------------------------
    -- object contains an unknown
    ------------------------------------------------------------------
    FUNCTION Is_X ( s :  std_ulogic_vector ) RETURN  BOOLEAN;
    FUNCTION Is_X ( s :  std_logic_vector  ) RETURN  BOOLEAN;
    FUNCTION Is_X ( s :  std_ulogic        ) RETURN  BOOLEAN;
END std_logic_1164;
```

B.4 THE STANDARD SYNTHESIS PACKAGES

The IEEE Standard synthesis packages include two VHDL packages: NUMERIC_BIT and NUMERIC_STD. The NUMERIC_BIT package is based on type BIT, while the NUMERIC_STD package is based on type STD_LOGIC.

B.4.1 NUMERIC_BIT

The package declaration of the NUMERIC_BIT is listed here for reference purpoe [4].

---

[4] Copyright ©1995 IEEE. All rights reserved. This is an unapproved IEEE Standards Draft, subject to change

Appendix B

```
--
--    Copyright 1995 by IEEE. All rights reserved.
--
--    This source file is considered by the IEEE to be an essential
--    part of the use of the standard 1076.3 and as such may be distributed
--    without change, except as permitted by the standard.
--    This source file may not be sold or distributed for profit.
--    This package may be modified to include additional data required
--    by tools, but must in no way change the external interfaces or
--    simulation behaviour of the description.  ie it is permissible to
--    add comments and/or attributes to the package, but not to change or
--    delete any original lines of the approved package.
--
--    Title      :   Standard VHDL Synthesis Package (1076.3, NUMERIC_BIT)
--
--    Library    :   This package shall be compiled into a library
--                   symbolically named IEEE.
--
--    Developers:    IEEE DASC Synthesis Working Group, PAR 1076.3
--
--    Purpose    :   This package defines numeric types and arithmetic functions
--               :   for use with synthesis tools.  Two numeric types are defined
--               :   -- > UNSIGNED: represents an UNSIGNED number in vector form
--               :   -- > SIGNED: represents a SIGNED number in vector form
--               :   The base element type is type BIT.
--               :   The leftmost bit is treated as the most significant bit.
--               :   Signed vectors are represented in two's complement form.
--               :   This package contains overloaded arithmetic operators on
--               :   the SIGNED and UNSIGNED types.  The package also contains
--               :   useful type conversions functions, clock detection
--               :   functions, and other utility functions.
--               :
--               :   If any argument to a function is a null array, a null array
is
--               :   returned (exceptions, if any, are noted individually).
--               :
```

Standard Library Packages

```
--
--      Limitation:
--
--      Note       :   No declarations or definitions shall be included in,
--                 :   or excluded from this package.  The "package declaration"
--                 :   defines the types, subtypes and declarations of
--                 :   NUMERIC_BIT.  The NUMERIC_BIT package body shall be
--                 :   considered the formal definition of the semantics of
--                 :   this package.  Tool developers may choose to implement
--                 :   the package body in the most efficient manner available
--                 :   to them.
--                 :
-- ----------------------------------------------------------------------
--  modification history :
-- ----------------------------------------------------------------------
--  Version:       INTERMEDIATE
--  Date   :       12 April 1995
-- ----------------------------------------------------------------------
package NUMERIC_BIT is
constant CopyRightNotice :String:="Copyright 1995 IEEE. All rights reserved.";
  --=========================================================================
  -- Numeric array type definitions
  --=========================================================================
  type UNSIGNED is array (NATURAL range <> ) of BIT;
  type SIGNED is array (NATURAL range <> ) of BIT;
  --=========================================================================
  -- Arithmetic Operators:
  --=========================================================================
  function "abs" (ARG: SIGNED) return SIGNED;
  function "-" (ARG: SIGNED) return SIGNED;
  --=========================================================================
  function "+" (L, R: UNSIGNED) return UNSIGNED;
  function "+" (L, R: SIGNED) return SIGNED;
  function "+" (L: UNSIGNED; R: NATURAL) return UNSIGNED;
  function "+" (L: NATURAL; R: UNSIGNED) return UNSIGNED;
```

```
function "+" (L: INTEGER; R: SIGNED) return SIGNED;
function "+" (L: SIGNED; R: INTEGER) return SIGNED;
--==============================================================================
function "-" (L, R: UNSIGNED) return UNSIGNED;
function "-" (L, R: SIGNED) return SIGNED;
function "-" (L: UNSIGNED; R: NATURAL) return UNSIGNED;
function "-" (L: NATURAL; R: UNSIGNED) return UNSIGNED;
function "-" (L: SIGNED; R: INTEGER) return SIGNED;
function "-" (L: INTEGER; R: SIGNED) return SIGNED;
--==============================================================================
function "*" (L, R: UNSIGNED) return UNSIGNED;
function "*" (L, R: SIGNED) return SIGNED;
function "*" (L: UNSIGNED; R: NATURAL) return UNSIGNED;
function "*" (L: NATURAL; R: UNSIGNED) return UNSIGNED;
function "*" (L: SIGNED; R: INTEGER) return SIGNED;
function "*" (L: INTEGER; R: SIGNED) return SIGNED;
--==============================================================================
function "/" (L, R: UNSIGNED) return UNSIGNED;
function "/" (L, R: SIGNED) return SIGNED;
function "/" (L: UNSIGNED; R: NATURAL) return UNSIGNED;
function "/" (L: NATURAL; R: UNSIGNED) return UNSIGNED;
function "/" (L: SIGNED; R: INTEGER) return SIGNED;
function "/" (L: INTEGER; R: SIGNED) return SIGNED;
--==============================================================================
function "rem" (L, R: UNSIGNED) return UNSIGNED;
function "rem" (L, R: SIGNED) return SIGNED;
function "rem" (L: UNSIGNED; R: NATURAL) return UNSIGNED;
function "rem" (L: NATURAL; R: UNSIGNED) return UNSIGNED;
function "rem" (L: SIGNED; R: INTEGER) return SIGNED;
function "rem" (L: INTEGER; R: SIGNED) return SIGNED;
--==============================================================================
function "mod" (L, R: UNSIGNED) return UNSIGNED;
function "mod" (L, R: SIGNED) return SIGNED;
function "mod" (L: UNSIGNED; R: NATURAL) return UNSIGNED;
```

Standard Library Packages

```
function "mod" (L: NATURAL; R: UNSIGNED) return UNSIGNED;
function "mod" (L: SIGNED; R: INTEGER) return SIGNED;
function "mod" (L: INTEGER; R: SIGNED) return SIGNED;
--=============================================================================
-- Comparison Operators
--=============================================================================
function ">" (L, R: UNSIGNED) return BOOLEAN;
function ">" (L, R: SIGNED) return BOOLEAN;
function ">" (L: NATURAL; R: UNSIGNED) return BOOLEAN;
function ">" (L: INTEGER; R: SIGNED) return BOOLEAN;
function ">" (L: UNSIGNED; R: NATURAL) return BOOLEAN;
function ">" (L: SIGNED; R: INTEGER) return BOOLEAN;
--=============================================================================
function "<" (L, R: UNSIGNED) return BOOLEAN;
function "<" (L, R: SIGNED) return BOOLEAN;
function "<" (L: NATURAL; R: UNSIGNED) return BOOLEAN;
function "<" (L: INTEGER; R: SIGNED) return BOOLEAN;
function "<" (L: UNSIGNED; R: NATURAL) return BOOLEAN;
function "<" (L: SIGNED; R: INTEGER) return BOOLEAN;
--=============================================================================
function "<=" (L, R: UNSIGNED) return BOOLEAN;
function "<=" (L, R: SIGNED) return BOOLEAN;
function "<=" (L: NATURAL; R: UNSIGNED) return BOOLEAN;
function "<=" (L: INTEGER; R: SIGNED) return BOOLEAN;
function "<=" (L: UNSIGNED; R: NATURAL) return BOOLEAN;
function "<=" (L: SIGNED; R: INTEGER) return BOOLEAN;
--=============================================================================
function ">=" (L, R: UNSIGNED) return BOOLEAN;
function ">=" (L, R: SIGNED) return BOOLEAN;
function ">=" (L: NATURAL; R: UNSIGNED) return BOOLEAN;
function ">=" (L: INTEGER; R: SIGNED) return BOOLEAN;
function ">=" (L: UNSIGNED; R: NATURAL) return BOOLEAN;
function ">=" (L: SIGNED; R: INTEGER) return BOOLEAN;
--=============================================================================
```

```vhdl
function "=" (L, R: UNSIGNED) return BOOLEAN;
function "=" (L, R: SIGNED) return BOOLEAN;
function "=" (L: NATURAL; R: UNSIGNED) return BOOLEAN;
function "=" (L: INTEGER; R: SIGNED) return BOOLEAN;
function "=" (L: UNSIGNED; R: NATURAL) return BOOLEAN;
function "=" (L: SIGNED; R: INTEGER) return BOOLEAN;
--============================================================================
function "/=" (L, R: UNSIGNED) return BOOLEAN;
function "/=" (L, R: SIGNED) return BOOLEAN;
function "/=" (L: NATURAL; R: UNSIGNED) return BOOLEAN;
function "/=" (L: INTEGER; R: SIGNED) return BOOLEAN;
function "/=" (L: UNSIGNED; R: NATURAL) return BOOLEAN;
function "/=" (L: SIGNED; R: INTEGER) return BOOLEAN;
--============================================================================
-- Shift and Rotate Functions
--============================================================================
function SHIFT_LEFT  (ARG: UNSIGNED; COUNT: NATURAL) return UNSIGNED;
function SHIFT_RIGHT (ARG: UNSIGNED; COUNT: NATURAL) return UNSIGNED;
function SHIFT_LEFT  (ARG: SIGNED; COUNT: NATURAL) return SIGNED;
function SHIFT_RIGHT (ARG: SIGNED; COUNT: NATURAL) return SIGNED;
--============================================================================
function ROTATE_LEFT  (ARG: UNSIGNED; COUNT: NATURAL) return UNSIGNED;
function ROTATE_RIGHT (ARG: UNSIGNED; COUNT: NATURAL) return UNSIGNED;
function ROTATE_LEFT  (ARG: SIGNED; COUNT: NATURAL) return SIGNED;
function ROTATE_RIGHT (ARG: SIGNED; COUNT: NATURAL) return SIGNED;
--============================================================================
-- ----------------------------------------------------------------------------
-- Note : This function is not compatible with VHDL 1076-1987.  Comment
-- out the function (declaration and body) for VHDL 1076-1987 compatibility.
-- ----------------------------------------------------------------------------
function "sll" (ARG: UNSIGNED; COUNT: INTEGER) return UNSIGNED;
function "sll" (ARG: SIGNED; COUNT: INTEGER) return SIGNED;
function "srl" (ARG: UNSIGNED; COUNT: INTEGER) return UNSIGNED;
function "srl" (ARG: SIGNED; COUNT: INTEGER) return SIGNED;
```

```
function "rol" (ARG: UNSIGNED; COUNT: INTEGER) return UNSIGNED;
function "rol" (ARG: SIGNED; COUNT: INTEGER) return SIGNED;
function "ror" (ARG: UNSIGNED; COUNT: INTEGER) return UNSIGNED;
function "ror" (ARG: SIGNED; COUNT: INTEGER) return SIGNED;
--=============================================================================
-- RESIZE Functions
--=============================================================================
function RESIZE (ARG: SIGNED; NEW_SIZE: NATURAL) return SIGNED;
function RESIZE (ARG: UNSIGNED; NEW_SIZE: NATURAL) return UNSIGNED;
--=============================================================================
-- Conversion Functions
--=============================================================================
function TO_INTEGER (ARG: UNSIGNED) return NATURAL;
function TO_INTEGER (ARG: SIGNED) return INTEGER;
function TO_UNSIGNED (ARG, SIZE: NATURAL) return UNSIGNED;
function TO_SIGNED (ARG: INTEGER; SIZE: NATURAL) return SIGNED;
--=============================================================================
-- Logical Operators
--=============================================================================
function "not" (L: UNSIGNED) return UNSIGNED;
function "and" (L, R: UNSIGNED) return UNSIGNED;
function "or"  (L, R: UNSIGNED) return UNSIGNED;
function "nand" (L, R: UNSIGNED) return UNSIGNED;
function "nor" (L, R: UNSIGNED) return UNSIGNED;
function "xor" (L, R: UNSIGNED) return UNSIGNED;
-- ------------------------------------------------------------------
-- Note : This function is not compatible with VHDL 1076-1987.  Comment
-- out the function (declaration and body) for VHDL 1076-1987 compatibility.
-- ------------------------------------------------------------------
function "xnor" (L, R: UNSIGNED) return UNSIGNED;
function "not" (L: SIGNED) return SIGNED;
function "and" (L, R: SIGNED) return SIGNED;
function "or"  (L, R: SIGNED) return SIGNED;
function "nand" (L, R: SIGNED) return SIGNED;
```

```
    function "nor" (L, R: SIGNED) return SIGNED;
    function "xor" (L, R: SIGNED) return SIGNED;
    --  ------------------------------------------------------------------
    --  Note : This function is not compatible with VHDL 1076-1987.  Comment
    --  out the function (declaration and body) for VHDL 1076-1987 compatibilit
    --  ------------------------------------------------------------------
    function "xnor" (L, R: SIGNED) return SIGNED;
    --==================================================================
    -- Edge Detection Functions
    --==================================================================
    function RISING_EDGE (signal S: BIT) return BOOLEAN;
    function FALLING_EDGE (signal S: BIT) return BOOLEAN;
end NUMERIC_BIT;
```

B.4.2 NUMERIC_STD

The package declaration of the NUMERIC_STD is listed here for reference purpose [5].

```
-- ------------------------------------------------------------------
--
--  Copyright 1995 by IEEE. All rights reserved.
--
--  This source file is considered by the IEEE to be an essential
--  part of the use of the standard 1076.3 and as such may be distributed
--  without change, except as permitted by the standard.
--  This source file may not be sold or distributed for profit.
--  This package may be modified to include additional data required
--  by tools, but must in no way change the external interfaces or
--  simulation behaviour of the description.  ie it is permissible to
--  add comments and/or attributes to the package, but not to change or
--  delete any original lines of the approved package.
--
--  Title    :      Standard VHDL Synthesis Package (1076.3, NUMERIC_STD)
```

---

[5]Copyright ©1995 IEEE. All rights reserved. This is an unapproved IEEE Standards Draft, subject to change

Standard Library Packages

```
--
--   Library    :    This package shall be compiled into a library
--                   symbolically named IEEE.
--
--   Developers:     IEEE DASC Synthesis Working Group, PAR 1076.3
--
--   Purpose    :    This package defines numeric types and arithmetic functions
--              :    for use with synthesis tools. Two numeric types are defined:
--              :    --> UNSIGNED: represents UNSIGNED number in vector form
--              :    --> SIGNED: represents a SIGNED number in vector form
--              :    The base element type is type STD_LOGIC.
--              :    The leftmost bit is treated as the most significant bit.
--              :    Signed vectors are represented in two's complement form.
--              :    This package contains overloaded arithmetic operators on
--              :    the SIGNED and UNSIGNED types. The package also contains
--              :    useful type conversions functions.
--              :
--              :    If any argument to a function is a null array, a null array is
--              :    returned (exceptions, if any, are noted individually).
--              :
--
--   Limitation:
--
--   Note       :    No declarations or definitions shall be included in,
--              :    or excluded from this package. The "package declaration"
--              :    defines the types, subtypes and declarations of
--              :    NUMERIC_STD. The NUMERIC_STD package body shall be
--              :    considered the formal definition of the semantics of
--              :    this package. Tool developers may choose to implement
--              :    the package body in the most efficient manner available
--              :    to them.
--              :
-- -----------------------------------------------------------------
--   modification history :
-- -----------------------------------------------------------------
```

```vhdl
--    Version:    INTERMEDIATE
--    Date   :    12 April 1995
-- ----------------------------------------------------------------------
library IEEE;
use IEEE.STD_LOGIC_1164.all;
package NUMERIC_STD is
constant CopyRightNotice :String:="Copyright 1995 IEEE. All rights reserved."
  --============================================================================
  -- Numeric array type definitions
  --============================================================================
  type UNSIGNED is array (NATURAL range <> ) of STD_LOGIC;
  type SIGNED is array (NATURAL range <> ) of STD_LOGIC;
  --============================================================================
  -- Arithmetic Operators:
  --============================================================================
  function "abs" (ARG: SIGNED) return SIGNED;
  function "-" (ARG: SIGNED) return SIGNED;
  --============================================================================
  function "+" (L, R: UNSIGNED) return UNSIGNED;
  function "+" (L, R: SIGNED) return SIGNED;
  function "+" (L: UNSIGNED; R: NATURAL) return UNSIGNED;
  function "+" (L: NATURAL; R: UNSIGNED) return UNSIGNED;
  function "+" (L: INTEGER; R: SIGNED) return SIGNED;
  function "+" (L: SIGNED; R: INTEGER) return SIGNED;
  --============================================================================
  function "-" (L, R: UNSIGNED) return UNSIGNED;
  function "-" (L, R: SIGNED) return SIGNED;
  function "-" (L: UNSIGNED;R: NATURAL) return UNSIGNED;
  function "-" (L: NATURAL; R: UNSIGNED) return UNSIGNED;
  function "-" (L: SIGNED; R: INTEGER) return SIGNED;
  function "-" (L: INTEGER; R: SIGNED) return SIGNED;
  --============================================================================
  function "*" (L, R: UNSIGNED) return UNSIGNED;
  function "*" (L, R: SIGNED) return SIGNED;
```

```
function "*" (L: UNSIGNED; R: NATURAL) return UNSIGNED;
function "*" (L: NATURAL; R: UNSIGNED) return UNSIGNED;
function "*" (L: SIGNED; R: INTEGER) return SIGNED;
function "*" (L: INTEGER; R: SIGNED) return SIGNED;
--=============================================================================
--
-- NOTE: If second argument is zero for "/" operator, a severity level
--       of ERROR is issued.
function "/" (L, R: UNSIGNED) return UNSIGNED;
function "/" (L, R: SIGNED) return SIGNED;
function "/" (L: UNSIGNED; R: NATURAL) return UNSIGNED;
function "/" (L: NATURAL; R: UNSIGNED) return UNSIGNED;
function "/" (L: SIGNED; R: INTEGER) return SIGNED;
function "/" (L: INTEGER; R: SIGNED) return SIGNED;
--=============================================================================
--
-- NOTE: If second argument is zero for "rem" operator, a severity level
--       of ERROR is issued.
function "rem" (L, R: UNSIGNED) return UNSIGNED;
function "rem" (L, R: SIGNED) return SIGNED;
function "rem" (L: UNSIGNED; R: NATURAL) return UNSIGNED;
function "rem" (L: NATURAL; R: UNSIGNED) return UNSIGNED;
function "rem" (L: SIGNED; R: INTEGER) return SIGNED;
function "rem" (L: INTEGER; R: SIGNED) return SIGNED;
--=============================================================================
--
-- NOTE: If second argument is zero for "mod" operator, a severity level
--       of ERROR is issued.
function "mod" (L, R: UNSIGNED) return UNSIGNED;
function "mod" (L, R: SIGNED) return SIGNED;
function "mod" (L: UNSIGNED; R: NATURAL) return UNSIGNED;
function "mod" (L: NATURAL; R: UNSIGNED) return UNSIGNED;
function "mod" (L: SIGNED; R: INTEGER) return SIGNED;
function "mod" (L: INTEGER; R: SIGNED) return SIGNED;
--=============================================================================
```

```
-- Comparison Operators
--==========================================================================
function ">" (L, R: UNSIGNED) return BOOLEAN;
function ">" (L, R: SIGNED) return BOOLEAN;
function ">" (L: NATURAL; R: UNSIGNED) return BOOLEAN;
function ">" (L: INTEGER; R: SIGNED) return BOOLEAN;
function ">" (L: UNSIGNED; R: NATURAL) return BOOLEAN;
function ">" (L: SIGNED; R: INTEGER) return BOOLEAN;
--==========================================================================
function "<" (L, R: UNSIGNED) return BOOLEAN;
function "<" (L, R: SIGNED) return BOOLEAN;
function "<" (L: NATURAL; R: UNSIGNED) return BOOLEAN;
function "<" (L: INTEGER; R: SIGNED) return BOOLEAN;
function "<" (L: UNSIGNED; R: NATURAL) return BOOLEAN;
function "<" (L: SIGNED; R: INTEGER) return BOOLEAN;
--==========================================================================
function "<=" (L, R: UNSIGNED) return BOOLEAN;
function "<=" (L, R: SIGNED) return BOOLEAN;
function "<=" (L: NATURAL; R: UNSIGNED) return BOOLEAN;
function "<=" (L: INTEGER; R: SIGNED) return BOOLEAN;
function "<=" (L: UNSIGNED; R: NATURAL) return BOOLEAN;
function "<=" (L: SIGNED; R: INTEGER) return BOOLEAN;
--==========================================================================
function ">=" (L, R: UNSIGNED) return BOOLEAN;
function ">=" (L, R: SIGNED) return BOOLEAN;
function ">=" (L: NATURAL; R: UNSIGNED) return BOOLEAN;
function ">=" (L: INTEGER; R: SIGNED) return BOOLEAN;
function ">=" (L: UNSIGNED; R: NATURAL) return BOOLEAN;
function ">=" (L: SIGNED; R: INTEGER) return BOOLEAN;
--==========================================================================
function "=" (L, R: UNSIGNED) return BOOLEAN;
function "=" (L, R: SIGNED) return BOOLEAN;
function "=" (L: NATURAL; R: UNSIGNED) return BOOLEAN;
function "=" (L: INTEGER; R: SIGNED) return BOOLEAN;
```

Standard Library Packages

```
function "=" (L: UNSIGNED; R: NATURAL) return BOOLEAN;
function "=" (L: SIGNED; R: INTEGER) return BOOLEAN;
--============================================================================
function "/=" (L, R: UNSIGNED) return BOOLEAN;
function "/=" (L, R: SIGNED) return BOOLEAN;
function "/=" (L: NATURAL; R: UNSIGNED) return BOOLEAN;
function "/=" (L: INTEGER; R: SIGNED) return BOOLEAN;
function "/=" (L: UNSIGNED; R: NATURAL) return BOOLEAN;
function "/=" (L: SIGNED; R: INTEGER) return BOOLEAN;
--============================================================================
-- Shift and Rotate Functions
--============================================================================
function SHIFT_LEFT (ARG: UNSIGNED; COUNT: NATURAL) return UNSIGNED;
function SHIFT_RIGHT (ARG: UNSIGNED; COUNT: NATURAL) return UNSIGNED;
function SHIFT_LEFT (ARG: SIGNED; COUNT: NATURAL) return SIGNED;
function SHIFT_RIGHT (ARG: SIGNED; COUNT: NATURAL) return SIGNED;
function ROTATE_LEFT (ARG: UNSIGNED; COUNT: NATURAL) return UNSIGNED;
function ROTATE_RIGHT (ARG: UNSIGNED; COUNT: NATURAL) return UNSIGNED;
function ROTATE_LEFT (ARG: SIGNED; COUNT: NATURAL) return SIGNED;
function ROTATE_RIGHT (ARG: SIGNED; COUNT: NATURAL) return SIGNED;
--============================================================================
-- ---------------------------------------------------------------------------
-- Note : This function is not compatible with VHDL 1076-1987.  Comment
-- out the function (declaration and body) for VHDL 1076-1987 compatibility.
-- ---------------------------------------------------------------------------
function "sll" (ARG: UNSIGNED; COUNT: INTEGER) return UNSIGNED;
function "sll" (ARG: SIGNED; COUNT: INTEGER) return SIGNED;
function "srl" (ARG: UNSIGNED; COUNT: INTEGER) return UNSIGNED;
function "srl" (ARG: SIGNED; COUNT: INTEGER) return SIGNED;
function "rol" (ARG: UNSIGNED; COUNT: INTEGER) return UNSIGNED;
function "rol" (ARG: SIGNED; COUNT: INTEGER) return SIGNED;
function "ror" (ARG: UNSIGNED; COUNT: INTEGER) return UNSIGNED;
function "ror" (ARG: SIGNED; COUNT: INTEGER) return SIGNED;
--============================================================================
--    RESIZE Functions
```

```
--=============================================================================
function RESIZE (ARG: SIGNED; NEW_SIZE: NATURAL) return SIGNED;
function RESIZE (ARG: UNSIGNED; NEW_SIZE: NATURAL) return UNSIGNED;
--=============================================================================
-- Conversion Functions
--=============================================================================
function TO_INTEGER (ARG: UNSIGNED) return NATURAL;
function TO_INTEGER (ARG: SIGNED) return INTEGER;
function TO_UNSIGNED (ARG, SIZE: NATURAL) return UNSIGNED;
function TO_SIGNED (ARG: INTEGER; SIZE: NATURAL) return SIGNED;
--=============================================================================
-- Logical Operators
--=============================================================================
function "not" (L: UNSIGNED) return UNSIGNED;
function "and" (L, R: UNSIGNED) return UNSIGNED;
function "or"  (L, R: UNSIGNED) return UNSIGNED;
function "nand" (L, R: UNSIGNED) return UNSIGNED;
function "nor" (L, R: UNSIGNED) return UNSIGNED;
function "xor" (L, R: UNSIGNED) return UNSIGNED;
function "not" (L: SIGNED) return SIGNED;
function "and" (L, R: SIGNED) return SIGNED;
function "or"  (L, R: SIGNED) return SIGNED;
function "nand" (L, R: SIGNED) return SIGNED;
function "nor" (L, R: SIGNED) return SIGNED;
function "xor" (L, R: SIGNED) return SIGNED;
--  ---------------------------------------------------------------------------
-- Note : This function is not compatible with VHDL 1076-1987. Comment
-- out the function (declaration and body) for VHDL 1076-1987 compatibility.
--  ---------------------------------------------------------------------------
function "xnor" (L, R: UNSIGNED) return UNSIGNED;
function "xnor" (L, R: SIGNED) return SIGNED;
--=============================================================================
-- Match Functions
--=============================================================================
```

```
function STD_MATCH (L, R: STD_ULOGIC) return BOOLEAN;
function STD_MATCH (L, R: UNSIGNED) return BOOLEAN;
function STD_MATCH (L, R: SIGNED) return BOOLEAN;
function STD_MATCH (L, R: STD_LOGIC_VECTOR) return BOOLEAN;
function STD_MATCH (L, R: STD_ULOGIC_VECTOR) return BOOLEAN;
--============================================================================
-- Translation Functions
--============================================================================
function TO_01 (S: UNSIGNED; XMAP: STD_LOGIC := '0') return UNSIGNED;
function TO_01 (S: SIGNED; XMAP: STD_LOGIC := '0') return SIGNED;
end NUMERIC_STD;
```

INDEX

A

Adding operators, 47
Aggregate, 52
Algorithmic level, 233
Alias, 51
Allocation, 247
And(logical operator), 46
Architecture design, 3
Architecture, 19
Array types, 44
Assertion statements, 66
Asynchronous Mealy machine, 159
Asynchronous reset directive, 256
Asynchronous reset, 191
Attributes, 51

B

Based literals, 49
Behavior Model, 22
Behavior style architecture, 8, 21
Behavior synthesis, 12
Bit clock generator, 283
BIT data type, 40
BIT_VECTOR data type, 41
Blackjack dealer, 293
Block statements, 80
BOOLEAN data type, 40
Booth's algorithm, 200
Buffer, 17
Bus system, 145

C

Case statements, 64
Cell model, 242

Cell-based design, 7
CHARACTER data type, 40
Character literals, 49
Combinational logic, 122
Combinational section, 134
Communicating Finite State
    Machines, 173
Component declaration, 101
Component instantiation
    statements, 102
Concurrent signal assignments, 77
Concurrent statements, 74
Conditional signal assignments, 78
Configuration specification, 107
Constant, 37
Constrained array type, 44
Constraints, 239
Counter, 141, 193

D

Data object, 37
Data path, 147
Data types, 38
Dataflow style architecture, 10, 26
Datapath and controller, 155
Decimal literals, 49
Default Binding, 107
Delta time, 25
Design library, 31
Design process, 1
Directives, 252
Divider operator, 48
Don't care directive, 261
Driver, 23
Driver, 62

Enabled flip-flop, 133

E

Entity declaration, 16
Enumeration literal, 39, 49
Enumeration types, 39
Event, 21
Event, 25
Exit statements, 70
Exponentiation operator, 48

F

Finite state machine with
    datapath, 169
Flip-flop inference, 127
Flip-flop with asynchronous
    inputs, 131
Flip-flop with synchronous inputs,
    129
Flip-flop, 185
FSMD level, 233
FSMD, 169
Full custom layout, 5
Function directive, 258
Functions, 85

G

Gate level, 233
GCD calculator, 196
Generic declaration, 16

H

Hardware description language, 1
HDL, 1
HDL-based design, 8
High impedance, 143

I

Identifier, 50
IEEE 1076-1987, 40
IEEE NUMERIC_BIT, 96
IEEE NUMERIC_STD, 96

IEEE Standard Logic Package, 94
IEEE Standard Synthesis
    Packages, 96
IEEE STD_LOGIC_1164, 94
If statements, 63
Ignore_begin directive, 256
Ignore_end directive, 256
Indexed name, 50
Inline expansion directive, 258
Integer data type, 40

L

Latch inference, 124
Library compiler, 242
Library unit, 31
Literals, 49
Logic synthesis, 12
Logical operators, 46
Loop statements, 67

M

Mealy model, 159, 163
Memories, 217
Memory model, 242
Mixed-level description, 233
Mixed-level design, 12
Mod (exponentiation operator), 48
Mode, 16
  buffer, 17
  in, 17
  inout, 17
  out, 17
Module function directive, 258
Module model, 242
Moore machines, 156
Multiplying operator, 48
Multiplying operators, 48

N

Nand(logical operator), 46
NATURAL subtype, 40
Neg_delay, 241